great tide rising

Toward Clarity & Moral Courage
in a Time of Climate Change

great tide rising

KATHLEEN DEAN MOORE

COUNTERPOINT
BERKELEY

Library of Congress Cataloging-in-Publication Data

Names: Moore, Kathleen Dean, author.
Title: Great tide rising : towards clarity and moral courage in a time of planetary change / Kathleen Dean Moore.
Description: Berkeley, CA : Counterpoint, 2016.
Identifiers: LCCN 2015035618 | ISBN 9781619026995 (hardback)
Subjects: LCSH: Nature—Effect of human beings on. | Climatic changes—Moral and ethical aspects. | Environmental protection—Moral and ethical aspects. | BISAC: NATURE / Environmental Conservation & Protection.
Classification: LCC GF75 .M665 2016 | DDC 179/.1--dc23
LC record available at http://lccn.loc.gov/2015035618

Cover design by Gerilyn Attebery
Interior design by Megan Jones Design

ISBN 978-1-61902-906-4

COUNTERPOINT
2560 Ninth Street, Suite 318
Berkeley, CA 94710
www.counterpointpress.com

Printed in the United States of America
Distributed by Publishers Group West

10 9 8 7 6 5 4 3 2 1

For Zoey, Theo, and Lem.

May the world always be safe for the laughter of children.

CONTENTS

PART III:
a call to witness

PART IV:
a call to act

INTRODUCTION:
learning to navigate amid loss

MARY EVELYN TUCKER
Yale University

WHEN WE HUMANS are lost we need directions or a map. Earlier humans had religious and cultural traditions to guide them. They provided, even imperfectly, some markers on the road—rites of passage to navigate the changing stages of life from birth to death. Those rites have become less effective as religious traditions have become somewhat diminished in the noise of info-tainment. We need new guideposts to find our way amid large scale changes such as climate change and loss of species. How do we live through these most challenging times?

Kathleen Dean Moore sets out such guideposts by raising critical and challenging questions. She is resolute in her pursuit of clear answers and relentless in her moral reasoning.

Clearly, the nourishing of the human spirit and imagination is what is at stake in our present moment—a spirit and imagination that is shriveling before our very eyes with an anguish and confusion that is heart rending. Kathleen bears witness. She sees how the human spirit and imagination are deeply entwined in the living forms around us. Their destruction is diminishing our capacity to dream and to hope. She worries. Without vibrant oceans and rivers, without lush quiet forests, without the movements and sounds of animals about us we will create a silence even larger than the silent spring Rachel Carson predicted. The silence will engulf us in the sound of our own lament. She weeps. This lament will not end soon for we are just beginning to write the eulogies, to sing the requiems, to plant the markers for life that is disappearing before our very eyes. And, she observes, this is true in areas that are urban and suburban, as well as rural and wild.

We are dwelling in a period of mass extinction and climate change. Loss is all around us. We are engulfed by it and at the same time we are nearly blind to it. Yet we feel in our bones some kind of unspeakable angst that will not leave us in the depths of night or even at daybreak when the birds greet the sunlight again. This crushing feeling of unstoppable destruction is holding us back from acknowledging our grief. Such loss of life demands not only mourning, but also recognition that we are in a huge historical whirlwind. Kathleen is aware that it will require every part of ourselves to find our way out. Are we like Job struggling to hear the call of life in the midst of the whirlwind of inexplicable loss? Or are we like Jacob wrestling with the angel of life in the presence of death? Or, she wonders, are we like Noah collecting and counting the animals, birds, reptiles, and amphibians that will pass through this hourglass of extinction with us?

We are groping, we are limping, we are struggling. But in this groping, in this dark night we are seeking to return to who we are. We are beings of Earth, she writes, who feel the mysterious rhythms of life unfolding. We sense this in the arc from sunrise to sunset, in the migrating patterns of birds and wild animals, in the call of whales in the depths of the oceans, in the leaping and twisting of animals and children at play, in the smell of spring soil appearing through winter's snow. All of it sings to us in the movement of seasons as the planet finds its way around the sun and back again. These rhythms, she hopes, will ground us anew in the Earth that has brought forth and sustained life for billions of years. The rhythms have changed, yes, with climate change and with extinction. We are being uprooted from predictable seasonal time, yet she dares to uncover ways forward. Deep time grounds us; planetary awareness encircles us as never before.

Rediscovering who we are. Finding our purpose as humans to enhance life, not diminish it. This is her endless prayer. In this we embrace our still evolving role as children of the Earth and as a planetary species. No longer are we citizens of nations alone but of the entire globe. Our allegiance, she suggests, is moving ever outward from family, society, state, continent, to the blue green planet that is home, and even beyond to the ethics of The Cosmos. There is no time for wavering. Rather, we are ready to move carefully and humbly into our mature role as a mutually-enhancing species. To do otherwise is to risk the destruction of life on the planet, she warns.

We are currently at sea. But we know somehow that our muddling through will be crucial to finding our way back home again. Patience, courage, and endless endurance are part of this process. Maybe humor, hope, and moral clarity will steer us through the whirlwind, she muses. The lights from the shore are there if we can only come up to the deck to see them. In a dark moonless night this requires a new kind of learning of currents, of winds, and of stars for navigation. Kathleen Dean Moore offers such learning to create a compass into the future.

PREFACE:

at low tide, watching the world go away

WE ARE WADING in rubber boots at the rim of the sea, my grandson and I. Behind us is a limestone ledge that shelters green anemones and limpets; in front of us, a bed of eelgrass laid flat by the receding tide. It's a silver day in Alaska—shining, shivering seas and clouds so low you feel you could bump your head. My grandson leans over to poke a graying starfish.

"This one is soft. That's means it's sick." This child is three years old and already he knows the signs of starfish wasting syndrome. He gives the sea star a last poke with his forefinger and stands to gaze around the cove.

"His mom is around here someplace," he says, wrinkling his brow and not finding her. "He's sick. He needs a mom." I think that is undoubtedly true.

Just last year, this cove was full of sea stars. We saw them in every damp crevice, heaps of them, the purple stars, *Pisaster ochraceus*, and mottled stars, *Evasterias troschelii*, not only purple but green, red, brown, orange. This year, we come across only three or two stars, here and there, splayed on the shingle. These that remain are wasting away too, a hideous process. Lesions form. The tissues around them decay, so the sea star flattens and falls apart. An arm may crawl away, but soon it too turns to mush. Around our boots, torn arms and the wispy scraps of wasted sea stars float on the incoming tide.

It's a catastrophe, among many on a planet growing sour and hot, and I am afraid for this small child.

Leaning over, he pries up a large rock. The bottom is plastered with baby sea stars, no bigger than his thumb, and they are firm to the touch.

"Not sick," he says, and looks up at me with a three-year-old's grin, which is the most winning, the most beautiful grin in the history of creation, a grin for the triumph of all the planet's babies. I'm uncertain about the prospects of the little ones. I imagine that these sea stars, smaller than a dime, are destined also to waste away to a lace of flesh that folds, refolds as small waves push it to shore, just as sea stars are dissolving all along the Pacific coast, Mexico to Alaska.

If only there were a mom around here who could shelter the young lives and comfort us all. But what would such a mother do? How could she bear the sadness? I can't think of anything worse for any parents than to feel helpless, as pieces of their child's world break off and quietly go away.

A statement of scientific consensus, led by Stanford scientists, has badly shaken me: Unless all nations take immediate action, by the time today's children are middle-aged, the life-support systems of the Earth will be irretrievably damaged. I am holding the hand of a small child in a yellow raincoat and orange bib overalls. His little boots have long ago filled with water. His hair is damp and smells of salt. And I am staring at my boots and thinking of what it could possibly mean to this child, to live on a planet whose life-supporting mechanisms have frayed and fallen apart.

He sucks in his breath. "Hey! Guys! Come close and look. Come close and look." Under a blade of rainbow kelp, he has found the red, orange-spiked, phallus-shaped sea animal called a sea cucumber. How beautiful it is, and how beautiful is the human impulse to be astonished.

But there's this. Yesterday, on a beach only two miles from this one, sea cucumbers by the hundreds washed up, dead. I'd never seen anything like that before. Gloriously colored animals sagging under the sudden weight of the world, they rolled in with the tidal detritus, tangled in seaweed and slime.

What does this mean for the next generations, all this dying? Can the human species thrive in a world where other species are disappearing, even as we watch? I just don't know. And what does it mean for us, the people of the present, who desperately care about the world?

People ask me, "Why do you try so hard to stop the fossil fuel industries that are overheating the oceans and the air? You are just one person, and the dying has already begun."

My only answer is this little guy in the yellow slicker, who is just now squatting to touch a pair of rock blennies that are flicking around the damp sand. "Look," he says. "This baby fish is still happy, and this one feels good too."

❖

HOW SHOULD I think about this? New research shows that the starfish wasting disease is most likely caused by a virus, and the virus is enabled by the increasingly warm and acidic seas, and the seas are warming and souring because they are absorbing carbon dioxide that is produced by burning fossil fuels. Somewhere there is a corporate committee that is deciding to expand the production and sales of fossil fuels, and at the same time to invest heavily in the U.S. Congress to kill legislation that would create alternative energy technologies. Somewhere, a man is deciding to trade the prospects of the next generations for the chance to increase the power and profits of his industry.

Economists say that the carbon catastrophe is an economic problem, and of course it is; it will cost trillions to restore or replace the functionality of ecosystems that climate change will destroy, if it's possible at all. Military experts write that climate change is a national security problem, and of course it is; famine and flooding will force people from their homes in waves of desperate refugees, and armies will mobilize for control of water. Technological wizards insist this is a technological problem, and of course it is; new ways to generate energy might free the world from culture-killing fossil fuels. Economic issue, national security issue, technological issue—it's all of these. But fundamentally, global warming is a moral problem, and it calls for a moral response.

To take what we need to support our profligate lives, and leave a ransacked and destabilized world for our grandchildren, is not worthy of us as moral beings. To let the world slip away—the starfish and sea anemones, the green and fecund marshland, the glacial streams—to let it slip away, because we're too busy, or too comfortable to change, is a sin against creation. And when a corporation, to further increase profits that are already unimaginably immense—when a corporation, as part of its business plan, knowingly destroys the conditions of flourishing life on Earth? That is moral monstrosity on a cosmic scale.

My response is moral outrage, an outrage as deep as the depth of my love for the future generations wading wide-eyed in the intertidal life, and for the life itself, the heaps and networks of glorious green and striving things. In this book, I want to make clear why it's flat-out wrong to wreck the world. And I want to become clear about what my responsibilities to the world and its hopeful, grinning children call me to do. We, all of us, are called to gather the moral clarity and courage to confront the changes that are already upon the world, even as the winds grow stronger, the waves grow steeper, and the courses forward are obscure, dangerous, and sometimes strangely beautiful.

Thich Nhat Hanh said, of the Buddha, "The real power . . . was that he had so much love. He saw people trapped in their notions of small separate self, feeling guilty or proud of that self, and he offered revolutionary teachings that resounded like a lion's roar, like a great rising tide, helping people to wake up and break free from the prison of ignorance." That's what the world must do now—summon from every voice the lion's roar, gather from the seven seas the great rising tide, to stop the final plunder and wreck of the world.

For ten years, I have been writing and speaking about the warming planet, the violent fluxes of wind and water, the fitful climate and acidifying seas, about habitat destruction and extinction, about the call to life. In the summers, I study and write in a cabin beside a tidal flat on Chichagof Island in Southeast Alaska. In the other months, I'm in the thick of things, speaking with community activists, interviewers, university workshops, radio shows, students, and symposia of all sorts about the moral urgency of the environmental emergencies. This book is a culmination of a decade of thinking and speaking about a moral response to the devastation of the planet's life-supporting systems.

By training and profession, I am a philosopher; my work is to think as clearly as I can about humanity's role on the planet and the moral responsibilities that come with that role, responsibilities to our own humanity, to fellow beings, to the future. By lineage and inclination, I am a naturalist and nature essayist; my work is to celebrate the glorious world—every purple starfish and barnacle—and to grieve its disappearance. By choice and good fortune, I am a mother and now a grandmother; my work is to push back against those who would, for the sake of profit, squander it all and leave only a plundered world for those who come next.

So this book weaves abstract reflection and felt experience, much the way a river weaves its current through the stones and sluices of its riverbed. Ideas expressed in philosophical discourse, and experience expressed in personal essays, together tell a story of why I cannot, will not, sit by and watch the world go away.

There is debate in Western philosophy as to whether moral decisions should be based on rational calculation or on the *moral sentiments*—emotions such as pity, generosity, love, and anger. The debate seems useless to me. We need both: clear, honest, razor-edged reasoning *and* the emotions against which we measure its conclusions. One might say that we need a conversation between mind and heart, reason and feeling, clarity and courage. Our challenges are too great to use only part of our capacities to address them. Our gifts are too great to waste or ignore. The peril of the world calls us to analyze, observe, test, speculate, and reason—and also to grieve, celebrate, imagine, regret, fear, abhor, hope, and act.

It matters to speak clearly of all this. Contemporary philosopher Charles Taylor wrote, "Articulacy has a moral point, not just in correcting what may be wrong views but also in making the force of an ideal that people are already living by more palpable, more vivid for them; and by making it more vivid, empowering them to live up to it in a fuller and more integral fashion." I think so too.

And so this book is written for an extensive and important audience: people who are trying to live by their ideals, an assemblage of clear-voiced people who care about the world. My purpose is to help them find words for their caring and reasons for their resolve.

So here is what I have been thinking and feeling as I stand in rubber boots in a sloshing tide, searching for a healthy starfish. How can one address the hard questions of a storm-threatened time: What should I do? Who should I be? How can one celebrate and love this glittering world, even as it becomes a sickened and dangerous thing? What can be said in response to the arrogance and illogic of those who would wreck the world? What are the words to say to people who deeply care, words that will help them move forward with new joy, courage, and integrity? How did human decisions create the climate emergencies, and how might new thinking take advantage of this last chance for civilization to start again and get it right this time? And this most important question: How can people come together in the one thing that has the power to change history—a great rising wave of moral outrage at the plunder and the wreckage, and an affirmation of a better way?

Because the looming environmental emergencies present a moral crisis, they call the citizens of the world to our best and highest humanity. Climate change is a violation of human rights on an unimaginable scale. It is a failure of reverence for all the lives—the creatures of land and sea that are striving to continue. It is a violation of justice, casting the burdens of corporate profligacy on those who will never enjoy its benefits. It is a betrayal of the children. It is also a chance, perhaps a last chance, to redeem the promise of humanity, the evolved awareness in a soft body that, from the time it draws its first breath, seeks to love and be loved.

Because they are a moral crisis, global warming and extinction are also a crisis of the imagination. The world can't fight its way out of this. We have to think our way out. This will require as great an exercise of the human imagination as the world has ever seen. The clever minds of the hominid will have to reinvent not just lightbulbs and electric cars. Rather, we will have to reinvent ourselves, we humans, thinking in entirely new ways about who we are and how we will preserve the best of our humanity as the world in which we evolved becomes something else entirely.

Finally, climate change is a crisis of character. We people of this decade, by some terrible chance, are the ones who are witness to the end of one world and the beginning of another. Maybe none of us would have chosen this time to be alive. The decisions we make in the next few years will decide if we will, or will not, redeem a just and thriving planet. As the world we know goes away, we are called to courage, not only to face down the coal trains and corporate bullies but to face into the wind, to keep moving forward on a course that has no predictable destination, across waters with hidden, changing shoals. We are called to integrity, to do what we believe is right, even if it has little prospect of success. And we are called to love—fiercely, maybe futilely, acting always in defense of what we love too much to lose.

As THE TIDE runs out, past the blue mussels to the great slabs of kelp, my grandson walks the high-tide line, looking for treasures. The tide has rolled a many-stranded rope from kelp stipes, eelgrass, gulls' wings, clamshells—whatever the sea has carried. He brings me a shiny green stone and a deer's femur, and then something I have never seen before.

It's a chalky white disk, as light and patterned as if it had been tatted. From a central disk about the size of a quarter, raised ridges radiate out like petals and then abruptly stop. I turn it over in my hand. In fact, the whole thing is about the size of my hand, and there are twenty-one of those ridges. Clearly organic. Radially symmetrical. Beautiful as a bone. The lacy exoskeleton of some astonishing new creature? My grandson touches it to his cheek. Then I know what it is, and I have to sit down on the sand.

This is the sun-dried central disk of a giant sunflower sea star (*Pycnopodia helianthoides*) after its twenty-one legs have broken off and crawled away to die, killed by starfish wasting disease. *What a glorious creature,* I think but do not say, *to have died such a death.* What an improbable creature it once was. Gooey and heavy, sometimes a meter across, stubbled with spines like the gray chin of an ancient man, the sunflower sea star creeps after prey on 128,000 tube feet, slithering maybe a meter every minute. When one is stressed, it can shed its legs, a process that emits a chemical that warns of danger. How shimmering with menace the very seawater must have been on the day this creature died.

It is important to say this: There are many things worth saving. No matter what happens, there are many things worth saving. The Earth is filled, it is populated, it is shivering, with lives of beauty and astonishment—what a child's hand holds, and the hand itself, and the reverence in the hand's careful holding. The fate of these lives is not a matter of indifference or of economic expediency alone. These lives are the irreplaceable, priceless consequence of the creativity of the universe over fourteen billion years. Now that humans have taken on the role of Earth-changer, we take on as well the responsibilities of celebration, protection, and ferocious love.

Kathleen Dean Moore
Chichagof Island, Alaska
2014

PART I

it's wrong to wreck the world

the power of moral affirmation

MY NEIGHBOR IS a practical man. "Look," he says to me, "if you want to call people to action on climate change, talk to them about what moves people to action—self-interest, money, and fear. Don't tell them it's wrong to wreck the world. Tell them it's stupid or expensive or dangerous. Tell them that unless this nation moves faster on solar energy, the Chinese are going to eat our lunch. Tell them that climate change will make the cost of California lettuce go through the roof. Scare the shit out of them if you have to: Tell them that the most likely way that global warming will end is when nuclear winter seizes the world—a long, cold night for the human prospect when nations drop atomic bombs in battles over water and food. But don't talk to people about morality. People don't like to be preached at. Ethics never changed anything, and it's a waste of time when time is short."

At that point, I really have my back up. I want to tell him what moral discourse is. I want to describe the power of moral affirmation to change history. It's true that morality has had a bad reputation lately. Blame it on TV preachers, if you want, shouting about sin. Blame it on the confusion of ethics with religion. Blame it on the effect of a press that would rather expose the titillating horror shows of private lives than engage in a discourse about what is right and what is good. Blame it on the viral pathology of this non sequitur: that because I have a right to believe whatever I want, whatever I believe is right. Poor ethics, struggling to be truly seen, when it has been so terribly disfigured.

First, let us distinguish between the morality of prohibition and the morality of affirmation. True, there's something repellent about an ethic based on prohibitions—thou shalt not, and if you do, which of course you will, you are bad, which of course you are. The morality of affirmation, on the other hand, is a soaring invitation to affirm what you believe is good and just and beautiful and right, and to align your life to those values. When my colleagues and I do public events about climate ethics, we gather people in small groups and ask them to address these questions: "What do you care about most? What would you die for? What would you be willing to spend your whole life taking care of?" Then we ask, "What follows from the fact that you hold these values? Values have consequences in the real world. If you value this more than anything else, what will you do—or never do? How might you make that value evident and real and powerful in your life?" That's the morality of affirmation.

Distinguish also between moralizing and moral reasoning. Nobody likes moralizing. But the civilized world depends on moral reasoning. The distinction is tricky. *Moralizing* is foisting your beliefs onto others, without offering any good reasons: "You should believe X. Period." *Pontificating* is similar. It's foisting your beliefs onto others, while offering a very bad reason—namely, your own assumed authority: "You should believe X because I believe X and I'm a moral savant." Both of them are very different from *witnessing* ("I believe X").

All of these are different from moral reasoning, which is an essential social and logical skill that this nation seems to have lost in all the shouting and piety. Moral reasoning is discourse in which people affirm what they think is true or good or right and then, the crucial step, back their claims with reasons. This is an invitation to respectful dialogue, to listening, and even to changing your mind. This is civil discourse, with its power to test beliefs against experience, your own and others', and revise and improve them.

Think of the eighteenth-century conversations about basic principles of the "rights of man" that flew through the streets and meeting halls of Europe. Think of the debates about life, liberty, and the moral responsibilities of revolution that took place in Philadelphia, even as they were taking place in the country towns. This was moral discourse, on every corner. Moral reasoning is that kind of conversation. "We can do this," I tell my neighbor. "We can talk reasonably about ethics."

And this claim that ethics never changed anything? I think that is a misreading of history. When you look at the times in American history when our society has turned on a dime, making huge social change, you'll find it was motivated by a great rising tide of affirmation of a strongly held moral principle.

No one says that moral affirmation is a sufficient condition for change, but I believe it is a necessary one. "We hold these truths to be self-evident, that all men are created equal"—a moral principle if there ever was one, and the old monarchies fell like dominoes from Europe across the Atlantic. "All persons held as slaves within any state . . . shall be then, thenceforward, and forever free"—a moral affirmation of the equality of all people, and the direction of history changed its flow. "I have a dream that one day this nation will rise up"—and the troopers and the growling dogs fell back. The nation is constantly involved in moral discourse, reasonable or not. So the question is not whether we should have a public discourse about ethics. The question is whether we can stop the ruin of Earth's ecosystems without the same affirmation of a widely held moral principle: "All beings have intrinsic value, and so have a right to a healthy and life-sustaining planet. And this right overrides the presumed right of the few to plunder the common heritage and destabilize the Earth's future without restraint."

And one more thing, I tell my neighbor: Ethics is a trump card, so it has a strong strategic value in social change—a value higher than that of the economic-gain card. "We can make a lot of money by enslaving our neighbor's children," one might say. The sentence that is sure to end that conversation is this: "But it would be wrong."

why it's wrong to wreck the world

SOME YEARS AGO, my colleague Michael P. Nelson and I were at a gathering of environmental philosophers in the ancient forests of the Oregon Cascade Range. It was raining, I'm sure of that; these five-hundred-year-old red cedars and Douglas firs are so tall that they catch on the clouds. Even if it wasn't raining from the sky, it was raining from the mats of moss and old man's beard that drape every branch and from the licorice ferns that grow in every crack and joint of the bigleaf maples.

Michael is different from most philosophers. He's funny and irreverent and worried about the world, and especially about wolves. He's philosopher-in-residence at the Isle Royale wolf study in Lake Superior and a professor in the College of Forestry at Oregon State.

We had gathered these people to bring good minds to the problems of the environmental crises but were frustrated almost to fury by the tendency of our colleagues to stand in one of the most heart-achingly beautiful places in the world and debate whether the green color, that vibrating, nourishing green, was present in the forest itself or merely in the eyes of us philosophers.

Michael and I drew away and pounded our heads to think of how we, as philosophers (or *qua* philosophers, as we philosophically said), could make a difference. Scientists did a heroic job of telling the public that climate change is real, it is dangerous, and it is upon us. Nothing happened. So they redoubled their efforts, speaking more loudly, in more resonant venues, even taking classes in communications, the dears, convinced that if people had sufficient understanding of the science, they would take action. Again, what happened? Not much. That, we realized, was because the scientists were making a basic logical error.

Consider the practical syllogism, brought to us by none other than Aristotle. Any argument that ends with a conclusion about what ought to be done is going to have to have two premises. The first is a factual premise, usually based on science: This is the way the world is, and this is the way the world is likely to be if civilizations continue on this course. But you can't reason from a claim about what is to a claim about what ought to be without a second premise. This is a normative premise, a premise that sets a standard: This is the way I believe the world ought to be. From these two, but from neither alone, you can discern what ought to be done.

So this is the point. Scientists have reached an overwhelming consensus about the facts of climate change. What is needed is a corresponding consensus about the moral urgency of climate action, backed up by a clear idea of our values. What we need, then, is a global conversation about the second premise: What do we dream? What do we cherish? What do we seek? What do we value? What is a good world? If we don't know this, we don't know where to aim our policies, or how to change our lives.

Michael and I set out to create a global conversation about the second premise. We wrote to one hundred of the world's truth-tellers—people such as Desmond Tutu, Ecumenical Patriarch Bartholomew I of Constantinople, Wangari Maathai, the Dalai Lama, Wendell Berry, N. Scott Momaday, and Sheila Watt-Cloutier—and asked them to answer a single question, in two thousand words or less: *Do we have a moral obligation to the future to leave a world as rich in possibilities as our own?* They sent back beautiful essays, full of conviction and light, and we published them in a book, *Moral Ground*. Although the reasons they gave were as many and varied as the soils or ice or sheep's-wool carpets on which these people walk, the reasons fell into a pattern woven from thirteen different threads. Michael and I summarized them this way:

THIRTEEN GOOD REASONS TO SAVE THE WORLD

First, think about the consequences that acting, or failing to act, will have on what we deeply value. And isn't this what it means to live a moral life—to do what you can to protect and nurture what you believe is good and beautiful and of great worth?

1. We must act for the sake of human life and thriving. This is as basic as it gets. If human life and thriving are fundamental values, and if climate disruption will have terrible costs in human lives and prospects, then we have a moral obligation to avert climate disruption. There are those who deny the value of human life, arguing that the Earth would be better off without the hominoid "plague," and there are those who deny the dangers of climate change. But if

you value humans and fear climate change, there is no excuse for not acting.

2. We must act for the sake of the children. It may be their innocence. It may be their promise—children holding the future of the human race in their little bodies. It may be our love for them. Or maybe it's just the driving force of evolution—the urge to leave something of ourselves to the world. For whatever reasons, humankind shares a universal moral imperative to prevent harm to children. But we are harming children, even as (especially as) we amass wealth to provide for them, destabilizing and denuding the world they will inherit. If we have an obligation to protect children from harm, and if climate disruption is manifestly harmful to them, then we have an obligation to expend extraordinary effort to prevent its worst effects.

3. We must act because all flourishing is mutual. Ecological science, ancient wisdom, and almost all the religions of the world tell us that life is interconnected. Humans are shaped and nourished by intricate relationships with air, climate, animals, soil, sun. Accordingly, human life is utterly dependent on the thriving of other beings and the stability of other systems—and anyone who professes to care about humans but not about "nature" is profoundly misguided. If we have an obligation to protect human thriving, then we have an obligation to protect the thriving of all the parts of the systems on which we depend—from the smallest ecosystems to the grandest workings of time and wind.

4. We must act for the sake of the Earth, its great systems, and its abundance of lives. The failure to do what we can to stabilize Earth's systems is, of course, a great imprudence—a cosmic cutting-off-the-limb-you're-standing-on stupidity. But it is also a moral failure. That is because the Earth as we have found it, a lonely

green jewel in the solar system, has value beyond its usefulness as the substrate of human lives. Sparrows and seagrass, newborn whales and tons of krill, fish on coral reefs, lingonberries and bears, all living things—could they have value only as they serve human needs? Only a colossal self-centeredness could lead one to think so. Rather, they have intrinsic value, value for their own sake. We have an obligation—even beyond our own interests—to protect a planet of inestimable and unique worth.

THE PRECEDING SET of four reasons to act on climate change is based on the consequences of acting, or failing to act. But we are also called to act honorably, regardless of the consequences. These are the reasons for acting that are grounded in our moral duties.

5. We must honor our duties as stewards of divine creation. "And it was good," God said of His creation. "And it was very good." The fish of the sea and the birds of the air and every living thing that creeps on Earth—all of these are good in the eyes of God. Imagine the fury and grief of the Creator, to see His creation trampled under the march of human greed. "To commit a crime against the natural world," Ecumenical Patriarch Bartholomew said, "is a sin." Religious institutions and their believers have a sacred responsibility to protect divine creation from what threatens it. And the greatest of these threats are human-caused climate and ecological disruption.

6. We must honor our duties to protect human rights. If all people have rights to life, liberty, and the pursuit of happiness, then the perpetrators of climate change are embarked on the greatest violation of human rights the world has ever seen. Millions will be denied the most basic of human rights as they are driven from their

homes by heat, drought, or flood; threatened by thirst, famine, and disease; and undermined by upheaval and war. Denial of human rights is a crime on any scale, in any legal system. So we all, corporations and governments especially, must end the policies that destabilize the climate—or be held to moral and legal account.

7. We must honor our duties to act justly. Those who are reaping the benefits of the profligate use of fossil fuels are casting off the terrible burden of their actions on those least likely to benefit and least able to defend themselves—future generations, poor and marginalized people everywhere, voiceless plants and animals. This is unjust.

8. We must honor our duties to future generations. The moral principle of "Do unto others as you would have others do unto you" applies with particular force in the face of the climate changes we have unleashed on the next generations. Our own lives rest securely on the acts and forbearance of our ancestors. Like us, our children will need clean air to breathe, clean water to drink, food to eat, room to stand, and natural beauty to lift and comfort them. These are the conditions of thriving that we minimally owe the children, and each of these is threatened by climate disruption.

9. We must honor our duties of gratitude and reciprocity. We did not earn this world. If it were taken away, there would be nothing we could do to get it back. The Earth, its life, and our lives are a gift. The gift calls us to defend and nurture the regenerative potential of the Earth, to return Earth's generosity with our own gifts of healing.

❖

As reasons one through nine make clear, moral action can stem from a sense of duty and from a calculation of the consequences of acting or not acting. But when all is said and done, moral action stems from human virtue. Who are we, when we are at our best? What actions grow from what is best in us?

10. We must act because we are compassionate. Compassion is the capacity to imagine oneself in another person's place, to feel their suffering as if it were one's own. But innocent suffering is the currency in which humankind will pay the price of the reckless use of fossil fuels—suffering from disrupted food supplies, degraded habitats, contaminated drinking water, infectious disease, terrific storms, destructive floods. A compassionate person will not allow the future to become a plague of sorrows.

11. We must act because we love the world. Like loving a person, loving a place is a way of feeling—joyous, connected, at peace. But that's not all. Loving is a sacred trust. To love is to affirm the absolute worth of what you love and to pledge your life to its thriving. A person who loves this world in all its beauty and comfort will not let it slip away through indifference or preoccupation, nor betray it for a passion for lesser things.

12. We must act because we feel the beauty of the world. However it came to be, the world is a beautiful creation—astonishing, wondrous, awe-inspiring at every scale. To ruin a dark shore's living planktonic lights, more glorious than any starry night Van Gogh could have painted, or to kill every tree on a mountain more awe-inspiring than Bierstadt could have imagined? This is not merely vandalism; this is a moral crime. A person who perceives the beauty of the world will protect it fiercely and faithfully.

13. We must act because we are people of integrity. There may come a time when hope fails us, and it seems that nothing we do will make a difference. There may come a time when a sense of duty

fades in the face of exhaustion. When that time comes, what's left to us is the power and the joy of moral integrity. To act with moral integrity is to match our actions to our moral beliefs, to do what's right, because we believe it's right, and for no more complicated reason. Through each everyday decision about what to invest in, what to eat, what to learn, what to buy and what to refuse to buy, how (or whether) to raise children, how to spend time, and how to treat neighbors, we can make our lives into works of art that embody our deepest values.

THIS LIST OF reasons to save the world doesn't seem like an academic exercise to me. Sometimes it seems like a prayer, whispered at the edge of a child's bed. At other times, it feels like the lyrics to a hip-hop song with crowds dancing and drumming; or maybe it's a call and response with two voices from opposite sides of an echoing canyon.

We must act for the sake of human life and thriving.
We must act for the sake of the children.
We must act for the mutual flourishing of all life.
We must act for the sake of the Earth.

We must honor our duties as stewards of divine creation.
We must honor our duties to protect human rights.
We must honor our duties to act justly.
We must honor our duties toward future generations.
We must honor our duties of gratitude and reciprocity.

We must act because we are compassionate.
We must act because we love the world.

We must act because we feel the beauty of the world.

We must act because we are people of integrity.

Not all of these reasons speak equally powerfully to all people, and that is as it should be. Different values motivate different people. Michael and I didn't set out to find the one overriding reason to act, but to collect all the reasons we could think of, somewhat on the legal principle of parallel pleading: In a life-or-death case, a wise defense lawyer will give the judge lots of different reasons to decide for her client. The flat fact is that, as Derrick Jensen puts it, this culture is not sustainable, and if we can't change its course, it will kill everything on the planet until it brings itself to its knees. The question then becomes, *What do you care about too much to lose?* The answer to that question is your reason for acting.

For myself, I come back again and again to the three reasons to act that are most likely to wake me, mind spinning and heart racing, in the dark hours before morning. Here they are, one by one.

because the world is wonderful

LET US BEGIN with wonder. Actually, let us begin with Wonder Bread. I don't remember what took me to the Oregon seacoast on the day I want to tell about. My memory begins as I was standing by my car on the south jetty on a gray day. I was watching sea lions dive for fish along a rock gabion that jutted into the channel. What had caught my attention was the small cloud that appeared each time a sea lion snorted to clear its nostrils. It was a measure of how cold the day was, and how hot and wet the sea lion's lungs must have been, and I was thinking about the miraculous transformation of cold wet fish flesh into hot wet mammal breath and flaring my own nostrils to catch the smell of the little clouds, when I heard a car slowly grinding gravel along the jetty road.

WONDER, BREAD

It was a white Buick, trailing a string of gulls. It parked beside me in a gravel pullout. Catching up the wind, the gulls winged furiously over the Buick, squawking. The passenger door opened on the far side of the car. Bedroom slippers on thin legs lowered themselves to the ground. Without warning, slices of bread flew up like toast from a cartoon toaster, and gulls swarmed to the open door, screaming and fighting.

On the driver's side, a woman opened the door, grasped the door frame, and pulled herself to standing. From her shoes to her blouse, she was dressed in lavender, and her hair, short and tightly permed, was brick red. The gulls circled her as she made her way to the back of the car. When she opened the trunk, the gulls went wild. Screeching, they swooped in close, colliding in midair. More bread popped up on the far side of the Buick. The gulls glanced over and dove for it, clattering their wings together, crying out.

The woman reached into her trunk for a loaf of bread. She unwound the twist tie and held it between her lips. Then she pulled out as much bread as she could hold in one hand. Gulls pressed against her legs, stumbled over her shoes. In spasms of excitement, they tilted back their heads and gulped out the raucous food call. The woman tossed up a handful of bread. Gulls caught the bread on the fly. What fell to the ground disappeared under slapping yellow feet and flapping wings. Gulls swooped in to pull at the bag the woman held in her hand. They swallowed quickly, and who could blame them, tossing back their heads and gulping down the scraps before another gull could snatch them.

How many gulls? A hundred? Two hundred? I sidled closer, not wanting to scare the birds or intrude on the old woman, but wanting to feel the wind of these flapping wings. She saw me coming.

"Want some?" she offered, and in fact, I really did. She beckoned me over to her trunk. Every niche was crammed with bread, one plastic grocery bag after another, each bag stuffed with five full loaves. This was soft, white Wonder Bread. When I was a kid, this was the bread we used to slather with margarine and coat with as much sugar as wouldn't shake off. It's a wonder we ever grew up.

"Safeway sells it," she said, although I hadn't asked. "Five loaves for a dollar."

"Good price," I said, because it was.

"We've been doing this every day for ten years. It's what we do."

I stood next to her and tossed bread into the wind. Knowing perfectly well that gulls couldn't live on white bread, knowing perfectly well that it deformed their wings, I fed it to them anyway. I threw a slice to a pure white gull that had only one eye, firmly fixed in my direction. I threw a slice to a gray-winged gull that had only one leg. But it didn't matter where I aimed; every bird mobbed every piece of bread. Birds hung at our heads, wings flapping and legs dangling. They swarmed at our feet. Feathers and bird droppings fell from the sky. Experimentally, I sidearmed three slices into the crowd. The volume of screaming was directly proportional to the amount of bread in the air.

Another handful of bread shot up from the far side of the car. A phalanx of gulls peeled off and settled by the bare feet in the bedroom slippers.

"My husband," the woman said.

Ah. "Do the birds follow you home?" I wanted to know.

"No. But they know we'll be back," she said, and turned to pull another loaf from her trunk.

She gave me fully half the loaf. I would have liked to have eaten it, I was that hungry for what the old woman offered—not just for the bread but for her closeness to the birds. But I tore the bread to pieces and threw it to the birds. Then I backed out of the melee. There stood the woman in her purple shoes, her face lifted to the birds, her arms wide open in the universal gesture of exaltation. Gulls fluttered around her like moths.

In the weeks after I met the gull lady, I thought about how one goes about living like that, with that extravagant joy and astonishment, how it becomes what you do, that hard embrace of what is wonderful, which is everything, when you think about it, every single thing in this mysterious, miraculous, morning-drenched world.

And her husband, the mechanical bread-throwing machine in the maroon bathrobe? He had fellowship, I would say. He had beautiful beings who flocked to him, the way his children probably once ran to him at the end of the day, the way his students (I'm guessing) once gathered around him, all eager hunger. Every day for ten years, faithfully, without fail, living beings sought him out for what he had to give them, even though he had so little left—except for the Wonder Bread, which he had in great abundance.

It fed us, the woman, the man, and me. That's what I want to say. Never mind the 140 calories, 180 milligrams of sodium, twenty-nine grams of carbohydrates, and two grams of dietary fiber in every two slices. It nourished us, we humans, to be surrounded by flocks of living beings of astonishing beauty and intelligence. (Astonishing, from the Latin word *tonus,* which means "thunder"—to be struck, as by lightning: the sudden flash that startles and, just for a moment, lights the world with uncommon clarity.) Humans need this delight, the way plants need sunlight. So we seek it out, going to the places where life is abundant, or bringing it into our homes, drawing it toward us with sunflower seeds and cracked corn. People by the thousands, sitting at their desks, link to webcams focused on endangered peregrine falcons who are feeding pigeons to their young in nests in high places: the Mid-Hudson

Bridge near Poughkeepsie, the Times Square Building in Rochester, the fourteenth floor of 55 Water Street in New York.

If there is a fact true of human beings—today, as always—I would suggest it is this: that we want to love and be loved, delight and be delighted, give and be given, in the back-and-forth relatedness that earns us a meaningful place in the pantheon of all being. The very muscles that allow us to raise our arms in gladness are the muscles that allow a gull to fly. I believe that this universal yearning, lifting toward life, is the greatest, most enduring, wonder of all.

Not so long ago, my daughter called from Tucson. Hearing a strange noise, she had come out the front door to investigate a sort of cracking and scratching noise, "as if somebody was trying to tear a mesquite tree apart with bare hands," she said. As she glanced around, she noticed that feathers were drifting from the sky and alighting on the gravel. She looked up. There, perched on top of a utility pole, was a sharp-shinned hawk, a mourning dove pinned in its talons. With its bony beak, the hawk pulled out the dove's feathers—that popping noise—and tossed them away. Each floating feather was soft and gray, with a puff of white at the end of the vane. My daughter stood in the yard, her arms spread, her face raised into a snowfall of bloody feathers. She was smiling, she said, but trying to keep her mouth closed so feathers wouldn't drift in.

"Souls that are focused and do not falter at first sight . . ," the philosopher Abraham Joshua Heschel wrote, "can behold the mountains as if they were gestures of exaltation." And here it was, on the jetty. The hovering gulls raised their heads and spread their cantilevered wings in the same gesture as my daughter's. Here was the bread and the tide in the channel and women in the wind of wings—everything beautiful and ineffable. Look! Look as if you have never seen this before, with that surprise, that wonderment. Look as if you would never see them again, with that yearning.

WHEN I LOOK with those new eyes at the story of the world, what I see rattles me, body and soul. From a single point in space and a single point in time—"a roaring force from one unknowable moment," Mary Evelyn Tucker calls it—all the elements in the universe burst forth. The elements self-organized into nebulae and stars and then galaxies and planets; in startling bursts of creativity, their patterns unfolded again and again. From those elements' unfolding came cells and the beginnings of self-replication and self-complexifying, life and lives developing like a fugue, variety unfolding, complexity unfurling, until, with the evolution of human consciousness, the generative urgency of the universe created a way to turn and contemplate itself. We are, she says, "beings in whom the universe shivers in wonder at itself."

Wonder, indeed, because here we are in the Cenozoic era, when evolution has achieved a great fullness of flowering, what theologian Thomas Berry called the most "lyric period in Earth history," the time of thrush-song and thirty thousand species of orchids, the time of microscopic sea angels with tiny wings and whales that teach one another to sing, the time of crocodiles and butterflies with curled tongues. Call out the names of the exquisite and roaring animals. Call out the names of spores and seeds.

I don't know if you think it was God who struck the downbeat that began this music, or if you think this glorious Earth is the result of the creative urgency of the universe alone. In a way, it doesn't matter. If you think it was God, do not think for a minute that He is indifferent. "And God saw that it was good, and it was good, it was very good," day after day, as the waters and the firmaments rained glistening life.

But say there is no Creator: that life created itself in great bursts of variation and selection, and filled the sky with midges and birds and filled the seas with not just fishes but the most extraordinary collection of creatures too various to be imagined—the creeping, chirping things with thin legs or sucking parts. If this is so, then this world is astonishing, irreplaceable, essential, beautiful and fearsome, generative, and beyond human understanding. If the good English word for this combination of characteristics is *sacred,* then that is the word I will use. We are born into a sacred world, and we ourselves are part of its glory.

This is the wonder-filled world that we are destroying, the lyric voices that we are silencing, the sanctity that we are defiling, at a rate and with a violence that cannot be measured because we have only the paltriest understanding of the world's multitudes of lives. Nonetheless, it's an extinction, scientists are able to agree, on the scale of the extinction that ended the Cretaceous period and the fern- and swamp-graced era of the stupendous dinosaurs. Then, an asteroid smashed into the Yucatán peninsula. Now, the destructive agent is human intention and disregard. This year, there are 40 percent fewer plants and animals on the planet than there were in 1974, when my daughter was born. In 2050, when her son is raising his own children, there will be 50 percent fewer species. His field guides will need only half as many pages, and the picture books about penguins and owls will be fantasies.

What does that matter? Why is it important that there be this planet with these odd little creatures? It could all end tomorrow. So what? Why should we care? We wouldn't know. Would anything of value be lost?

The answer of course is yes. It matters that a hundred years from now, salmon return to the streams, children hum themselves to sleep, red-legged frogs burble underwater.

We are struggling to talk about something of deep sacredness, anthropologist Mary Catherine Bateson says. The creativity of this living world is continuing to unfold. And that unfolding is sacred. Be prepared, she goes on to say, be prepared to wonder at this unfurling. Be prepared also for this: that every extinction, every suffering, every destruction, is a diminishment of creativity, and so it is a profanity. Be prepared for anger and for grief. The world is a mystery of infinite and intrinsic value. Be prepared to love it in ways beyond our own understanding. This wondering love is what brings us to the work ahead of us and sustains us in the struggle.

Why is it wrong to wreck the world? Because the world is a wonder, beautiful and creative, unique and irreplaceable. And what is wonderful ought to be honored and protected. The failure to honor and protect it is a failure of reverence.

Philosophers put this somewhat differently and debate it endlessly, but the point is the same. Two kinds of value can be distinguished, Immanuel Kant said centuries ago: instrumental value and intrinsic value. Some things have value as means to the ends of others. Some things have value as ends in themselves. A shovel, for example, is instrumentally valuable if you want to dig dirt. But in itself, apart from its usefulness, it is not worth much except as an example of human invention.

So what of the world? Does it have instrumental value? Absolutely, yes. It provides food, water, sanitation, warmth, shelter, beauty—all the necessary and sufficient conditions for human life. Economists call these "ecosystem services" and run the numbers about the replacement value of these services until their computers overheat. But something with instrumental value can also have intrinsic value. For example, my husband has great instrumental value. He is a wonderful provider; he fixes things that are broken, builds what needs to be built, and provides any number of pleasures. But he also has intrinsic value. His life is something of great worth in itself, apart from its usefulness as a means to my ends, and if I should forget that for any length of time, I will soon find out that his "services" are in fact "gifts" that can be readily taken away.

This planet, with all its lively systems, has instrumental value, beyond doubt. But it also has intrinsic value. Surely it is good—in itself.

To be clear about that, try what has come to be called the "last man" thought experiment. If you were the last person on Earth, leaving the planet in the last spaceship, never to return, would you, as you fastened your seat belt, press the hypothetical button that would destroy everything on the planet and leave it ruined and smoking in the rearview mirror? If the world has value only as it is useful to humankind, there would be nothing wrong with this. Nobody will return to the planet; it is suddenly useless. But if the world has intrinsic value, then pressing that destructive button would be wrong—the wanton destruction of something that is of value, even when human uses have rocketed away.

because we love the children

MY LITTLE GRANDDAUGHTER, Zoey, sings herself to sleep at night and laughs in her dreams. One snowy evening, I lay beside her and her beloved toy penguin while she hummed in the dark. She was humming a good song: *Laugh, kookaburra, laugh, kookaburra, gay your life must be.* The snow falling through the night, the small and trusting child in the blankets, a giant kingfisher guffawing on the other side of the globe—it was too much for me. "That's it," I said to myself. "It's over. From this moment on, I will do nothing in my working life except try to make the world safe for the laughter of children. And kookaburras."

If we fail to prevent the ruin of the life-support systems of the planet, if we strip the planet of its green and growing beauty, we will have betrayed our love for the children. The children did not create this danger. They do not deserve the struggle that is thundering toward them. Nonetheless, they will have to live in whatever is left of the world after the bulldozers and drilling rigs and toxic spills get done with it.

Why is it wrong to wreck the world? Because we love the children. "Because we swore and vowed to every god we ever imagined or invented or dimly sensed that we would care for them with every iota of our energy when they came to us miraculously from the sea of the stars," Portland poet Brian Doyle wrote. "Because they are the very definition of innocent, and every single blow and shout and shiver of fear that rains down upon them is utterly undeserved and unfair and unwarranted. Because we used to be them, and we remember, dimly, what it was like to be small and frightened and confused."

And, I would add, because we promised them. We held our newborn children and promised, "I will always love you. I will always keep you safe. I will give you the world."

Just last night, my husband showed me a cartoon from the *New Yorker*. There is the usual office scene—a woman in a suit behind a desk, and an eager old man bent into a chair. "I'd like to take out one of those mortgages on my grandchildren's future," he says. I laughed. I really laughed. No one would ever set out to do that, or only a monster would. But this is exactly what this dead-end culture is doing—maybe not intentionally, but knowingly. What will I say when my granddaughter comes to me with her own baby in her arms and real pain in her voice and asks me, "What did you do to protect the Earth from this devastation?" I cringe when I imagine what she might say:

Don't tell me you didn't know. You knew.

Don't tell me you thought there was enough time. You knew there wasn't.

Don't tell me you didn't know what to do. Anything would have been better than nothing.

Don't tell me the forces against you were too great. Nothing is greater than the forces against us now. And now, what would you have me do?

As a parent, I feel the obligation to children so strongly that I struggle to question it as a philosopher. What is it about little children, I asked in *Moral Ground,* that makes us grant them unconditional moral worth?

It may be their innocence. A small child can never deserve to suffer, because she can never do the wrong that might merit suffering in return. It may be their promise. Children contain the future of the whole human race in their little bodies. Anyone who values the future of humanity must therefore do right by children. It may be a parent's love for her child. Responsibilities grow from the relation of parent to child. As parents acknowledge the responsibility to care for their own children, they acknowledge the responsibility to care for all children, who can't be distinguished on moral grounds. Or maybe it's just the driving force of evolution. The protectiveness that humans feel toward children may be the manifestation of the urge to leave something of themselves in those who will come next. Regardless of what ethical theory one turns to, there will be a moral imperative to prevent harm to children. But what that means is complicated.

A tall man came up to me after one of my talks and stood only inches away. I couldn't see his face, so I don't know his expression; all I could see was the third button on his shirt. "I love my daughter more than anything else in the world," he said. "All I want is for her to be safe and happy. So I am going to amass as much money as I possibly can to provide for her, and I don't care how I get it. Is that so wrong?"

I would like to have pressed my forehead against that white button and wept in his arms. So sad. So sad, that we parents are harming children, even as (especially as) he and I believe we are acting to provide for them. Think of the privileged children: the poison in the plastic car seat, the disease in the pesticide-treated fruit, the disastrous coal company in the college investment portfolio, the mall where there had been frogs, the carbon load of the soccer tournament. Even as we try to help the children, these decisions harm their futures, stealing from them a world as rich and delightful as the world we were born to. It's a tragic irony that the amassing of material wealth we do in the name of our children's futures hurts them most of all.

But it's not just an irony; it's a moral abomination, what our decisions will do to children who are not privileged. These children, in distant countries and the distant future, will never know even the short-term benefits of misusing fossil fuels. But they are the ones who will suffer as seas rise, as fires scorch cropland, as diseases spread north, as famine comes to lands that had been abundant. And then all the children will be in it together on one reeling, keening Earth.

If parents have a moral obligation to protect the children, I would tell that tall man, and if environmental harms and climate change are manifestly harmful to them, then we have a moral obligation to expend extraordinary effort to immediately stop those harms and redress the wrongs that we have already done in their names.

What shall we give the children? Sandhill cranes—surely sandhill cranes. And the sweet whistle of the varied thrush in the morning. Frog calls, owl calls, trumpeting whales. Fresh cold water to drink at the end of a saltwater day. Deep green shade. Starfish, and a child's delight in these. Blueberries and potatoes. Safe nights. A sense of decency and fairness that will last them all their lives. Farsighted love.

"Think," I would say to the tall man. "Think about what your daughter really needs to thrive." In our desperate and single-minded love for our children, in the cacophony of dunning voices that do not have our children's interests in mind, we are like albatross parents, who dutifully bring their chicks bright, shining things they find on the seas—but not the nourishing squid and sardines, not anymore. What transoceanic commerce now offers the albatross are shards of plastic, bullet cases, broken toys, plastic beads—shiny as squid but empty of anything that would sustain life. When the parents return from foraging far and wide, this is what they regurgitate to nourish their chicks. Albatross chicks are dying in the Midway Islands, starving with their stomachs full.

"And our daughters?" I would say to the man. "What are they really hungry for?"

For that man—that loving father—and for the albatross, I have written this prayer. I imagine him speaking first and the albatross answering.

EVERY PARENT'S PRAYER

I will hold you in my arms and sing to you, a soft lullaby. *With my beak, I will nudge you into the sweet feathers of my chest and sing to you, a whistling song. I will keep you safe and defend you steadfastly. I will peck the eyes from anyone who tries to hurt you.*

You are the hoping of the universe, its wondering mind. You are the reason God made the fishes of the sea and the birds of the air and red apples and olives on trees. You are the child I have dreamed of forever and ever. Every night I have dreamed of you.

For you, the Earth created wing feathers, wind, and wild flight. You are the reason for the gleam on the krill. You are the reason for the slick sardines. You are my reason for living. I want the world for you. Yes, I will bring you the world and lay it at your feet. The world. I promise.

I will soar across wave tops, days and long nights, to bring you rainbow-tipped squid from the green seas. I will give you a rainbow ribbon for your hair.

I will bring you bloodied herring, broken by feeding whales. I will give you cherries in a yellow pail, and a pink pony with blue glass eyes.

I will bring you yellow stomachs and blue hearts tossed from fishing ships, oily eulachon and flying fish, caught as they leap from the sea. I will give you a plastic dolly with yellow hair, an emerald crown for her pretty head, and tiny slippers for her feet.

I will sail among the hooked lines of the fishing ships, I will trail the sharks, I will ride the tails of squalls to bring you what I find in their bleeding wakes—silver bodies, stunned or dead. I will buy you a plastic safety seat for the car.

Anchovies, flicking in a sapphire sea. For your high chair, a harness with a plastic clip.

A periwinkle and a nautilus. Rubber suction cups to keep your plate from sliding to the floor.

A purple kelp crab. Three more pairs of tiny shoes.

An octopus blushing pink, a pink shrimp. A Styrofoam bowl of ice cream with a red plastic spoon. *A yellow damselfish, end over end in a breaking wave.* A purple comb for the pony's mane. *A crab with purple stripes.* A red shovel. *Rain.* Calm seas. *Lively seas.* A button shaped like a puppy.

And when you spread your wings, beloved fledgling, may the wind lift you and hold you aloft. May the wind hold you. *May the seas feed you.* May the seas feed you. *May the Earth be kind to your fragile wings.* May the Earth be kind. *May your soaring flight answer every parent's prayer for his small child.* May your soaring flight answer every parent's prayer for his small child.

❖

I WOULD LIKE to tell a story now, a story about an angel and a duck, because I don't know any other way to say how complicated this is—this love, this fear.

THE ANGEL AND THE DUCK

It was Monday of the week before Christmas, and three generations of our family set out in two cars to cut a Christmas tree in the old fields we own outside of town. One might imagine snow and red knit caps and flocks of winter birds following the family parade across the field, the grandfather carrying the crosscut saw. This was not to be. Imagine instead a string of people hunkered under yellow raincoats, wading in high boots through dead grasses battered down by weeks of rain. But my husband, Frank, did carry the crosscut saw, and our son, Jonathan, carried Zoey on his shoulders, and you can imagine us laughing as we sloshed from tree to tree to find the perfect shape. This was to be the first Christmas Zoey would remember, and we wanted to do it right. So there she sat, holding on to Jon's head, resplendent in a hooded yellow rain slicker.

We hadn't thought this through.

"This one, everybody. It's perfect. Isn't it beautiful, Zoey?"

"Do you like this one, Zoey?"

"Tree," she said.

We took this as an affirmation. Agreeable and bewildered, as a child often is, she reached out to grab the tree and laughed when water splashed her face. When Frank trimmed off the bottom branches, she laughed and hugged the fronds. A child is as ready to love a tree branch as anything else, with an embracing, indiscriminate love. It took only a couple of strokes with the crosscut before the tree was wobbling on its stump. Then Frank gathered Zoey to his side, and with her little blue mitten next to his big glove, they made the last cut.

"Timber," we all yelled. Of course we did. That's what you do when a tree falls.

Zoey watched astounded as the tree swished down, fell on its side, bounced once, and rolled a short distance down the hill. Then she burst into tears. Of course she did. We had just brought down a tree, moments after pronouncing it the best tree in the whole world, ever. As I say, we hadn't really thought this through.

It wasn't long before the Douglas fir wobbled in the bay window, still fragrant with the spice of the forest. Boxes of ornaments littered the floor. Laughing and bending over the work, every adult in the family was trying to untangle strings of Christmas lights. As for Zoey, she was practicing the skill of walking backward while beeping like a truck. She backed into a cardboard box. Reaching out small hands, she pulled open the flaps. There, resplendent in tissue paper, lay an angel—blue dress, golden halo of curls, and magnificent outspread white-feathered wings. Zoey lifted the angel with both hands.

"Duck!" she said.

She brought the angel's dainty nose close to her own.

"*Quack!*" she said.

From that moment, the child and the angel could not be separated. She carried the angel everywhere, crushed it to her chest, fed it Cheerios, and patted its golden hair. Like any mother duck, she crouched on the floor, tucked the angel under her chest, and lay there with her eyes closed and her arms spread, sheltering the angel. At nightfall, her mother wrapped a blanket around the angel and the child and carried them both to bed.

By morning, glittering gifts would pile under the tree, but in my mind, those gifts would be inconsequential. The gift that I wanted for my granddaughter was the soft shelter of what might be Earth, but could well be mistaken for heaven. I wanted her to know the iridescence of the duck and its tail-up feeding on a flustered lake. I wanted her to know the lake itself, the green smell of it, its coolness at dusk, and the dark forest that encircled it, the hush of approaching rain, the last light that brightened a bead of water at the end of every branch. I wanted her to be safe in the world, and I wanted the world to be safe in her arms.

At night, Zoey cried out, awakening the household. A soft light glowed around the edges of the bedroom door, and I heard her mother speak quietly. "Don't be afraid," she murmured. "There's no reason to be afraid."

But I knew that on this deep and starless night, the whole world was awake and afraid. The child's fears were the world's night terrors. Under a half moon, cattle licked dust in the desert. Bedrock dissolved in the acid sea. Blue ice fell at the ends of the Earth. Saltwater snicked over seawalls. We grown-ups had pronounced the world good, perfect in every detail, and then we had severed it from its roots and hauled it away. Maybe we had already twisted the great swirling skies into storms that would change the world forever. I curled my body on the tangled sheets and tried to go back to sleep.

I didn't want to think about any of this. I wanted to fall into a dreamless sleep and awaken to the crowing of a small child who pads into my bedroom, holding an angel by the hair. I didn't want to think about this especially: that in some tragic confusion, we were wounding the world for the sake of the children. I wanted to wake up laughing and get out of bed and put on my bathrobe and, leading my little granddaughter by the hand, flip the switch that brought the tree alive with light and dazzle. But I couldn't escape this: that we grown-ups are ransacking the world where our children and grandchildren will have to live. There would be no sleep in the face of this.

The night, as it turned out, had been a tough one for the angel too. Her head gazed blankly backward and her golden toupee flapped over one ear. Her feathers were ruffled. Her silken dress, unsurprisingly, looked as though it had been slept in. This seemed not to matter to the angel and not at all to Zoey, who held the angel out to the see the lighted tree. The angel must have been impressed, because she flapped up and down in Zoey's hands, dragging a broken wing.

In a child's eyes, everything is equally miracle, ducks and angels. The good, green Earth is full of mysteries and sacred wonders, if we could see them as a child does—the duck's eagerness for flight, the damp morning, the call of a heavenly host of returning crows. For the sake of the children, we are called to sacrifice, which means literally this: that we are called to make the world sacred again. We are called to see that the light that glows behind the trees at the end of the day is a miraculous light, and it's the same light that catches in a child's hair, and startles a hatchling crow, and flares from burning methane gas. The world is not a lump of coal. The world is a great and astonishing improbability, a mystery of great beauty that has come, for reasons we will never understand, as a gift to us.

Behind a pile of crumpled newspaper and open cartons of decorations was another box. Jonathan pulled open a flap, and Zoey trundled over to help. She pulled out a framed photograph of my grandmother, which had been clipped from the front page of the Cleveland *Plain Dealer*. My son carefully positioned the picture in its traditional place on a shelf beside the tree. In the photograph, my grandmother is a small child, so the picture must have been taken more than a century ago.

We see the child from behind. She is wearing neat buttoned boots and a little fur-trimmed coat. Her head is thrown back. Curls spill over her shoulders. Her arms are lifted and her gloved hands are turned upward in the universal gesture of astonishment and joy. What has she seen? The doors of Macy's department store have just opened to reveal a towering Christmas tree that glitters, for the first time in the history of the town, with electric lights.

A century—my grandmother, my father, me, my son, my granddaughter—that fast; that's all the time it took. Here was one child who stood in rapture before the glory of a new technology. Now, here was another child who stood at the edge of a desecrated world, where even the angels' wings were torn beyond repair. I didn't know what to make of it. I sat in my living room, in my grandmother's chair, my grandmother's blue velvet chair, with a tangle of Christmas lights in my lap. One by one, I twisted the dead lights until I twisted one last light and the whole tangle sprang into color. I jumped, as if it were something alive.

So we draped the lights on the tree and we served the same roast beef and Yorkshire pudding dinner that my grandmother had served us. And then it was night again. Rain thudded on the roof. Yellow firelight pulsed across the darkened room. The only other light was from the spangled Christmas tree, red and blue and green. Zoey and the angel sagged in my lap. It was a silent night, a holy night, as all nights are holy. I felt . . . I'm trying to remember how I felt, exactly.

I felt as though I were saying good-bye to something, as if small angels were slipping away through the floorboards. I listened for any change, listened as an act of will, willing my legs to remember the pressure of Zoey's small body, willing my eyes to remember the silhouette of Jonathan's face against the fire. Somehow I knew that I would go over those details in my mind again and again. I felt I had to prepare for it all to be gone.

because we honor human rights and justice

IF THERE IS any moral principle on which the people of the world agree, it is that all people have the right to life, liberty, and security of person. The affirmation is found in international documents such as the Universal Declaration of Human Rights and the American Declaration of the Rights and Duties of Man, and in federal constitutions around the world, from Russia to the United States to South Africa. That said, if there is anything that threatens a more massive violation of human rights than runaway climate change, I cannot imagine what it would be, except perhaps for global nuclear war. The violation is on that scale.

Human rights. Sheila Watt-Cloutier, the former international president of the Inuit Circumpolar Council, lives in Nunavut, a vast Canadian landscape of ice sheets and islands. Wearing a fur-ruffed coat to ward off the winter, she spoke about the damage that global warming has done to her homeland, listing "eroded landscapes, coastal losses because of erosion, longer sea-ice-free seasons, the melting permafrost which is now causing beach slumping, increased snowfall in some areas, not enough snow in other areas, unpredictable sea-ice conditions." She continued: "Glaciers are melting, creating rivers instead of streams, and so more drownings where hunters thought they could cross safely. And so it is starting to undermine the ecosystem, the very land, ice, and snow that we depend on for our own physical and cultural survival." All this was factual observation, the first premise. But then she added the essential second premise, by affirming the moral right to her lifeways.

"As our culture is based on the cold, the ice and snow," she said, "we are in essence defending our right to be cold."

In 2005, Watt-Cloutier and a group of Inuit hunters took their case to the Inter-American Commission on Human Rights. The argument? That greenhouse gas emissions from the United States were undermining the material basis of the culture on which their lives, their liberties, and their security depended. That undermining was a violation of human rights, which was wrong, both morally and legally, and so it must end.

Americans are not unfamiliar with the injustice of destroying people and cultures by destroying their means of subsistence. Slaughter the bison to destroy the bison-based culture of the Plains Indians. Decimate the salmon to reduce the Salmon People to penury. Drain the swamps to drive out the Indians and escaped slaves. Blast the jungles with Agent Orange to defeat a nation.

The United Nations Office of the High Commissioner for Human Rights's report on climate change and human rights is quick to point out that the realization of human rights depends to a large extent on a healthy environment. The *right to life?* Climate change threatens human life directly and indirectly, through extreme heat, more severe storms and floods, higher winds, increased diseases, drought and desertification, and displacement caused by inundation, hunger, thirst, and war. The UN report points out as well that climate change is a direct threat to the *right to liberty,* what it calls "self-determination." People who are deprived of their means of subsistence no longer enjoy the right to shape their lives by their own choices. As for the right to what is variously called the *pursuit of happiness* or *security of person,* this too is stripped away when the life-supporting systems of a culture are degraded or destroyed.

One way to think of the magnitude of the violation of human rights is to imagine the world's response if it were not climate change caused by the industrial growth economy but an alien force that was responsible for undermining the life-support systems of the Earth. Imagine, Derrick Jensen urges, that aliens invade Earth. (Derrick is northern California's radically imaginative coauthor of *Deep Green Resistance.*)

Here the aliens are, swarming Earth, infiltrating everything. They set about to sicken children with asthma, poison the groundwater, stir the storms into devastating forces, pour oil over the shrimp beds. They inject poisonous wastes into caves until the Earth trembles. Summoning hurricanes, they drown old ladies in their attics. They pump acid into the oceans. They blast cropland into deserts. They inundate coastal cities. They kill off half the plants and animals on the planet.

What would we think? What would we do? (So, okay: At first we might not think much about it, denying any connection between the aliens and the devastation, or convincing ourselves that the aliens were actually working for the common good, or thinking with some satisfaction that the worst was far away and that the first victims were people we didn't really care about anyway.) But very soon, we would realize that this damage was a terrible threat to our fundamental rights to life, liberty, and security of person. At that point, I do think, we would go to tragic war. We would come together in common cause to expel the aliens from the planet, in hopes that we might protect our human rights, the fundamental decency of our lives, and the prospects of our children.

Justice. Around midnight on a clear night last spring, I was sitting in a window seat in a 737, flying the red-eye home to Portland. Over the intercom, the pilot told us to look out the left side of the airplane. We were thirty-three thousand feet over the Bakken oil fields in North Dakota. I had never seen anything like it in my life. The whole plain, horizon to horizon, was studded with flames. *How will I ever describe this?* I wondered. *With what analogy to something the human mind can grasp?*

It was like those pictures of the lava plains of early Earth, with all the rocks leaking flame. It was like a marching army of flames. Or this, more precisely: Think of one of the many heartbreaking World War I military cemeteries with closely ranked rows of crosses lining the hills as far as you can see in any direction. You think, *My god, how is it possible that humans would do this to one another?* Now imagine that you're in that cemetery, and it's night, and all those crosses have burst into flame. Flames, flaring off the methane from wellheads in closely ranked rows of fire as far as you can see in every direction, all the way to the curve of the horizon. You think, *My god, how is this possible? What is this doing to Earth and its creatures, and by what right?*

It's not just that people are suffering and will suffer from the looming carbon catastrophe, an assault on their human rights. The moral problem is that they will suffer unjustly, in violation of any principle of equitable distribution. The industrialized countries, which have contributed the most to climate change, are largely not the ones who are bearing its costs; the costs are disproportionately cast onto those who are least to blame.

The same injustice is occurring at the level of the person. Some people will enjoy the presumed advantages of reckless consumption of fuels and consumer goods, and in the process, some very few people will become obscenely rich. They will do so by offloading the damage of the profligate burning of fossil fuels on those who have not benefited from fossil fuels and who are least able to speak in their own defense. Who are they? Plants and animals. Children. Future generations. Poor and marginalized people—people on the raw edge of economies and on the muddy edge of continents. That's not fair.

Why not? What is justice? What is an equitable distribution of benefits and burdens? How does this relate to fossil fuels and climate change, habitat destruction and extinction, all the means and consequences of wrecking the world?

To answer, we will have to begin (of course, we will) in early Greece, when a just distribution was considered to be an equal one; in fact, Aristotle used the word for equality, *isos,* to refer to fairness. But you just try to give all the students in a class a C, and they will very quickly abandon the notion that justice can be achieved by equality alone. What they are looking for is proportionate equality—what we now call equity. A just distribution is one in which people receive the benefits and burdens that they deserve.

In a perfectly just world, the consequences of climate change would not fall equally on one and all. The burdens would fall squarely on the heads of those who are most benefited by its easy energy and most responsible for its harms. Those who are not responsible for climate change would escape its effects. But this is not a perfectly just world—not even close. In this world, it's exactly the opposite of ideal: The *burdens* of climate change—hunger and thirst, poisoned air and water, inundation, disruption, and wars—are imposed disproportionately on the world's poorest communities and those that are the least responsible for its effects.

Anywhere you look, you find examples. The glacial ice on the Tibetan plateau waters ten major river systems that provide irrigation, power, and drinking water for more than a billion people. When the glaciers are gone, the rivers will lose their reliable replenishment. What did those downstream people do to deserve that? One out of seven people on Earth depend on food from the sea. When the acidic seawater dissolves the tissue-thin phytoplankton that support the food pyramid, that food source will be gone. What did the people do to deserve that? Cancer flows from fracking wells and belches from the stacks of oil refineries. What did the sick children do to deserve that?

Scientists project that by the end of the century, drought will make 99 percent of Africa unsuitable for agriculture. How will the starving people deserve that? Nicholas Stern, an economist and climate change expert, says, "Hundreds of millions of people will be forced to leave their homelands because their crops and animals will have died. The trouble will come when they try to migrate into new lands, however. That will bring them into armed conflict with people already living there." These innocent beings will pay the deferred costs of runaway plundering—dear god, the injustice of their fate. And what could future people have possibly done to deserve an unstable and dangerous world?

So what about the unequal distribution of the *benefits* of the practices that cause climate change, counting among them huge profits for investors, breathtaking salaries for captains of industry, and an ease and abundance of life that the world has never seen before and will surely never see again? By what accounting do we (you and I and the barons) deserve this privilege? Are we perhaps more virtuous? That's tough to argue when profits are based on raiding the commons—the atmosphere, the soil, the forests, the freshwater, the mountaintops, the carboniferous past, the prospects of the future. Are we perhaps more enterprising? That's tough to argue when around the world fossil fuel industries are supported by subsidies at the rate of at least 775 billion to a trillion dollars per year. Are we luckier? Undoubtedly, yes, but that's not a matter of justice.

(And, parenthetically, our luck may soon run out. Climatologists draw pictures of two overlapping bell curves. Because of the properties of the ocean, the effects of global warming are being felt first in the southern hemisphere and the extreme north. But in a decade or so, the rising line of the second bell curve will begin to accelerate, and the effects will be increasingly felt in the northern temperate zones as the heat, drought, storms, and insects move north.)

So what is a fair distribution of benefits and burdens? The late Harvard philosopher John Rawls thought of it this way: A practice is fair if it is a practice that people would unanimously choose if they were blind to how that practice would affect them particularly. In other words, how would you arrange the world, by what rules, if you did not know in advance if you were going to be lucky or unlucky, black or white, sick or strong, male or female, Bangladeshi or Canadian? Under those circumstances, Rawls is quite sure that rational people would arrange the world in a way that would protect their own interests, even if they turned out to be in the least advantaged position. And so, he said, they would choose a practice that would treat everyone equally—or if some inequalities would be beneficial to all, the privileged places would be open to everyone.

Every professor I know compares it to cutting a child's birthday cake. If a mom wants to be sure of a fair distribution of cake, she will ask one child to cut the cake and give another child first chance to choose her piece. Because the child doing the cutting does not know what piece he will get, the mother knows that never in the history of birthday parties will there be a more meticulously equal division of pieces.

It's a revealing thought experiment: Equality is what happens when the people who decide how to cut the cake (senators, for example) can't rig the division to favor themselves.

But how can I figure out my fair share, or yours, of greenhouse gas emissions? Australian philosopher Peter Singer suggests this: Ask scientists to calculate the quantity of greenhouse gas emissions that the atmosphere can absorb each year without raising Earth's temperature more than two degrees Celsius. Take that number, then divide it by the number of people on Earth. That answer is each person's fair share of the carbon-absorbing capacity of the atmosphere.

Singer's answer presupposes that no one has a right to pollute more than anyone else. But when I compare his number to my existing annual carbon footprint, I learn that I am emitting about seven times as much carbon as I can justify by the principles of fairness. Clearly, I have work to do if I don't believe in taking more than my fair share.

WHATEVER IS LEFT of the human enterprise when climate change exhausts itself is what the new world will make itself from—just as what was left of the Cretaceous flora and fauna after the dust cleared was what the present world was made from. The cosmic challenge of our time is to preserve our humanity through this great crisis—not just our species, but our full humanity—so that the dignity and moral ideals of justice and the rights of humankind remain to flower into something strong and lasting.

PART II

a call to care

a love story

THEY'LL WANT TO know the story of this time, our grandchildren will, this pivotal decade when we either found our way forward or did not. They will give our time a name; it is that important. The Hinge Decade. The Last Blessed Time. The End Times. The Great Turning. The Eighth Day of Creation. "Was it like this," they will say, "storm clouds building on the horizon, desperate people pushing across borders, shouting and gunfire? Or was it quiet: simple disappearances, lost opportunities, lost species, quiet as snow melting, quiet as desiccation? Was it exciting: new ideas, a new sense of empowerment, a weepy joy of relief and redemption?"

They will wonder if we saw it coming. They will want to know how hard we really tried to stop the destruction. They will want an accounting. They will want a book filled with the prayers people prayed. They will want jars filled with the last of things.

The point is that we are the ones who will create the story of our times. What will it be? The story of the pivotal years of global warming could be a crime novel. It could be a horror story, with zombie cockroaches. It could be scripture, with all its terror and grace. It could be—no doubt *will be*—a thriller with a thousand plot twists. It could be an absurdist, nihilist farce.

Of course, we're living out the end of an old story right now. It's a story that has driven the industrial capitalist extractive economy for centuries—a worldview, a cosmology, a set of assumptions about who we are and what we ought to aspire to. For the last five hundred years or so, Western civilization has been living in an adolescent superhero comic book, the sort of cartoon fantasy of planetary subjugation and mastery that stirs the loins of teenage boys and Wall Street bankers.

The fantasy would have us believe that humans are the superheroes of the planet, so different from and superior to the rest of the Earthlings that we might have descended from the sky; the fantasy would have us believe that we are in charge of the planet, in control, wresting riches from the Earth, which lies supine and stupid at our feet.

The fantasy tells us that we are lonely heroes on a dangerous planet, in competition with one another. That competition has winners and losers—that's the way of the world—and the losers should be grateful to live on the toxic trickle-down excess of the heroes. Of course, superheroes are exempt from the rules that govern the rest of the world and the dumb brutes.

Even God is on the side of the superhero humans; He gave us this world. But who needs God anymore? If we get into danger, our superhero technologies will save us. Who needs fresh air when you have oxygen bars on the street corners and cigarette filters to stick up your nostrils (as people are doing now in China)? Who needs predictable weather when you live in a techno-bubble? Who needs other people to care for you when you have money enough to hire caregivers? Who needs glaciers when you have desalination plants? Who needs Earth when you have Mars? Who needs compassion when our superpower . . . ? Our superpower is the iron grip of Adam Smith's invisible hand—this magical hand that supposedly turns uncontrolled selfishness and greed into biocultural thriving.

That story is over. It is a failed experiment. The world doesn't work that way. We can't, in fact, destroy our habitat and one another without destroying ourselves. Based on an outdated scientific view of a mechanistic Earth, the old worldview is wildly inconsistent with emerging ecological and evolutionary understanding of an interconnected planet. Playing out that script has brought the entire planet to a cosmically dangerous place.

What's the next story? So far, it's playing out as a farce. Maybe you have read *Waiting for Godot* by Samuel Beckett, where two ragged men stand by a dead tree and wait. For whom is not clear. For what is never clear. I wrote a few lines of dialogue for our time:

What you waiting for, Vladimir?

Not waiting.

Yes you are.

Am what?

Waiting.

Oh.

Nothing I can do, just one person. Kinda sad about the polar bears.

Sad.

Congress might do something.

Might. Might not.

Something big might happen. A big storm.

Yeh.

But what's the use. You know: China.

Yeah.

Need proof.

I was thinking that.

Did I hear something?

No.

We've got to come up with a better story than this. Whatever it is, the new story has to have as much power as the story of creation—a story that is so powerful that it can bring itself into existence. "In the beginning" was the great creative unfolding, petals unfurling, and it was good, the tiger lilies and the walrus and the laughing children. And now, on the eighth day of creation, how can we imagine a story of a new creation, another great unfolding of new ideas and new lifeways, a new sense of ourselves in relation to the world, a blossoming?

The only story I know that has that kind of power will be a love story.

There are many kinds of love stories. There's the sad kind; everybody's seen this movie a hundred times. He doesn't know he loves her. Of course, we in the audience know she's spectacular and he's a dumbass not to notice, but he doesn't. Now and then, he begins to show signs of falling in love (his gaze lingers, his fingers linger), and we start to think there's hope for a happy ending. But of course there isn't. At some point, knowing he will never fully commit to a life-changing love, she throws her clothes into a suitcase and leaves. That's the exact moment when he realizes he can't live without her, and then there's the usual chase. She gets in a cab. The last we see of her is that long, silky leg folding into the backseat. He commandeers a car and goes careening after, running red lights, bouncing off buses, sprinting across the airport plaza with his necktie flying, and he gets to the gate just as the wheels of her plane leave the ground.

In many ways, the story of our time is starting to sound like that sad story. It's the oldest plot in history. Still, we have trouble understanding the storyline, even as we play our parts, even as we recite our lines. Will the hapless heroes understand how much they love this life on Earth before it all slips away? Will they learn in the stumbling ways of all lovers how to care for it? Or will this story end tragically, leaving only a bittersweet lesson, as actors and audience alike walk out of the theater into a rainy night, holding on to the only truth they can salvage from the sorrow—that love is a good and powerful thing, and even if what you love vanishes in the end, it is no less worthy of love and the love is no less beautiful. That's a tearjerker if I ever heard one.

But we live in a choose-your-own-adventure book. That's the point. The heart-thumper in this story is whether the hero will wake up to his love before it's too late. When will he ask himself, *What do I love too much to lose?* That's the point when his love becomes elemental and fierce, and his life becomes a story about the absolute determination that what he loves will not be allowed to pass away. What is called for is action that is the moral equivalent to the power of his love. At that point, he becomes the hero of his own story.

So now we set about to write the next chapter of the world. How do you write a story so huge that it has to call everything into question: how we feed ourselves, how we educate ourselves, how we exchange goods and services, how we share land, how we work, how we die?

Every writer knows the answer to this question. You write a story by writing it. Sentence by sentence. Scene by scene. Here's writer Anne Lamott answering that question:

> Thirty years ago my older brother, who was ten years old at the time, was trying to get a report on birds written that he'd had three months to write. It was due the next day . . . he was at the kitchen table close to tears, surrounded by binder paper and pencils and unopened books on birds, immobilized by the hugeness of the task ahead. Then my father sat down beside him, put his arm around my brother's shoulder, and said, "Bird by bird, buddy. Just take it bird by bird."

That's how we will take it. No one knows how the story will end. It is the work of a lifetime, many lifetimes. Every chapter is the context for the next. We will start with the setting. That would be the Earth, spinning. We will start with a protagonist. That would be us. We are the heroes of this journey. We are Bilbo Baggins and Odysseus and Harriet Tubman. We will start with a motivation. That would be fierce love. And then we will begin to make the story up. Even if it doesn't have a happy ending, the depth of our love of the planet and all its small lives, of our children, of justice and human rights, will be illuminated by the story unfolding. A story, like love itself, is not "a thing, after all, but an endless series of single acts."

BECAUSE I HAD been thinking about what it means—really means—to love a place, I carried a worry with me when I traveled north to be a writer-in-residence for a week in the heart of Denali National Park, where global warming has taken the land in its teeth. This is the question: What is the relation between an open heart and a broken heart? I mean, I love this land, its ice and grizzlies, its blue mountain peaks, even the sound of its wind. I know how desperately it is endangered. So shall I open my heart to this kind of love and risk its being broken? And because I often try to write my way into a philosophical problem, this is the small love story I wrote about this land:

AT THE EAST FORK CABIN, ALL IS WELL

"All is well at the East Fork Cabin," I shout at the satellite phone. Every day, at 7:00 AM and 7:00 PM, I am supposed to call in to the communication center at Denali National Park. If I don't, they will send a ranger over to make sure I haven't been routed by bears or nudged by a bus off Polychrome Pass. I am a writer-in-residence at the park, living in a cabin deep in wild country, and they need to take care of me.

"I read you," Randy the dispatcher shouts back. "All is well at the East Fork Cabin. Talk to you tonight." He signs off.

"No, really," I want to tell him. "Really, all is really well." I want to tell him that a bell chorus of water drops had sung all the night long, dripping into the creek from the ceiling of the snow bridge. The music was silenced by the cold of the next morning—just a *pock* here and a *pock* there from the arch over the water. In the early morning air, each willow catkin glistened with frost, and the snowdrifts over the creek bed were hard and shining. Ice sheets grew across the shallows, drawing silver topographic maps. Under the bank, crystals raised glass cathedrals with mud roofs. On the bank, ice filled the perfect paw print of a wolf.

Clicking off the phone, I take my seat on the bench by the river in a grove of aspen trees. Sometimes, a morning is so quiet that you can hear the breath of time itself, the slow in and out that shimmers in new leaves. Now, as the silted current sifts over stones, it rustles like an orchestra getting ready to play. The flute players puff air through the narrow tubes of their flutes to warm silver and brass. There it is—the soft brush hush as violinists lean forward to adjust their scores. That little *tap-tap*: a chickadee opens a seed and a percussionist tunes the timpani. Softly, a Swainson's thrush whistles up the scale.

Do you know the sound when all the members of a choir stand, the rustle of their rising? That's what the river sounds like, every pebble pushing up its wave. The road grader on the bridge beeps a backup warning. In the sudden silence, everything is poised to begin. The morning draws in its breath. Here now is the first flooding chord of water over pebble and cobble and boulder, and the basso profundo of the first bus downshifting on the grade over the pass. A ptarmigan cackles. Siskins whisper. Silt rasps against rocks, and cobbles roll. So indistinct but so musical, so full, the sounds I hear could be the Mormon Tabernacle Choir carried on the wind from yesterday, or from a thousand miles away.

Icy catkins are candles in this early light. Their flames flicker. Amber light flows through new poplar leaves. The northern anemones are still sleeping, or maybe they are praying, standing with their heads bent over, the petals closed across their faces. When I passed them last evening, their heads were thrown back like tenors' heads, full open to the sun. But now, the cells on one side of their stems have lost their timbre, and their heads nod.

And Randy, I want to tell you this—that when my father was dying, he listened over and over to the chords of the old hymn that ends with a great surge of voices and joyous trumpets: *All is well. All is well.* That's the refrain he listened to, reaching out his hand to press REPLAY. *All is well.*

I didn't understand it then, that a person can feel trusting love flow over him like water and reassurance fill him like music, even as his heart is failing. But maybe I am beginning to understand, because I've listened to a river carry a mountaintop to the sea. I've seen how darkness changes to light and back again, and mud to ice. The snow will melt, the white anemones will fade back into the Earth, the wolves will pass into the stars, our fathers' ashes will float to high valleys. But even when all lives are gone, there will still be the music of water on stone, and the faraway singing of the wind.

❖

THAT EXPERIENCE IN Denali confirmed what I have long believed, that humans can fall in love with places, much as they fall in love with people. Sometimes we fall in love with the entirety of places, the global home so beautiful from space, with its glowing blue skin and soft clouds. But I think that more often it's a particular place that makes us feel whole and happy and alive—the crest of a mountain, the childhood shelter behind the hedge, the rock-wrack shore of the sea, the lilac in bloom by the bus stop.

What does it mean to love a place? Is it different from loving a person? I decided I would make two lists: one list of what it means to love a place, and another list of what it means to love a person. Before long, I discovered that I was writing two copies of the same list. To love—a place or a person—means at least this:

1. To want to be near it, physically.

2. To want to know everything about it—its story, its moods, what it looks like by moonlight.

3. To rejoice in the fact of it.

4. To be transformed in its presence—lifted, lighter on your feet, transparent, open to everything beautiful and new.

5. To want to be joined with it, taken in by it, lost in it.

6. To fear its loss and grieve for its injuries.

7. To protect it—fiercely, mindlessly, futilely, and maybe tragically, but to be helpless to do otherwise.

My granddaughter, then in kindergarten, told me that I had left something off the list. Have I? "Yes, kissing," she said. "If you love something, you kiss it." I will go along with that:

8. To press your lips against it, to taste it, to close your eyes and feel it gratefully and fully.

I leaned down to kiss the top of her head. I felt her soft hair and smelled shampoo and cedar trees, because we had been lying on our backs in the Alaskan forest, watching sandhill cranes kettle in the white sky and listening to them call. They sounded like trumpets underwater. They sounded like ravens speaking German. Zoey laughed and called out to these astonishing birds, who were flying in on the north wind, as they have flown for nine million years. There is, as Rumi wrote, more than one way to kneel and kiss the ground.

on joyous attention

YES. *There is more than one way to kneel and kiss the ground.* Is paying close and joyous attention one of these ways? The temptation is to turn away from a sadly degraded world, but I'm starting to understand how an attitude of attentiveness to the natural world can be a matter of moral significance—that it may in fact be a keystone virtue in a time of reckless destruction, a source of decency and hope and restraint.

Soon I'll try to parse this argument. But first, a story:

BEAR SIGN

To get to our Alaskan cabin from the cove where we moor our boat, Frank and I walk a trail known as Bear Alley. Bears have used this path for so many hundreds of years that they have worn it into a narrow, flat-bottomed channel—a foot wide, maybe, a foot deep in the duff and moss, winding up the mountainside through lady ferns and Sitka spruce.

There are many more bears than people on this Alaskan island, and at least a couple of bears follow these paths each day. From our bunk at the window in the cabin, I often wake up to see a bear grazing on beach sedge across the cove, big and calm as a cow.

Am I afraid of the bears? Yes, in fact, I am. These are Alaskan brown bears, seacoast grizzlies grown huge and sleek on salmon. I want to live my allotted time, not find myself prematurely swatted into eternity because I got myself in the way of a bear. It doesn't matter that never, in the memory of this inlet, has anyone been so much as bruised by a bear; when I walk up the trail (usually carrying a five-gallon pail of halibut or crabs) or walk down the trail (usually carrying a bucket of fish carcasses or crab shells), I am fiercely attentive.

That circle of beaten-down grass—how long ago did a bear bed down there? That pile of bear scat like Ping-Pong balls woven of grass—has that been here since spring, when bears headed from their dens to the beach for a first meal? That patch of ground pines popping out of the duff—bears love them. If a bear had passed recently, wouldn't all the ground pines be uprooted and gnawed? Have hemlock needles fallen on that pile of scat? Has rain softened the edges of the paw print in the mud? I stop to examine the bears' scratching post at the intersection of the trails. They stop here and scratch their shoulders against the trunk. The bark on the tree is rubbed to a fine polish. I look for new patches of grizzled fur stuck in the cracks.

I scan the forest constantly, alert for a brown hump in the salmonberries. I listen, turning my head. Disturbed bears don't growl. They huff. Quiet as the huffing is, it is easily mistaken for water against stones. But when it comes from behind a screen of alder trees, it is a sound that catches one's attention. Sometimes, if I'm approaching a rise or a dense patch of berries, I'll let out a *hoo-eee,* as if I were calling pigs. But most of the time on the trail, I sing. I sing the songs my mother taught me, whatever old campfire song comes to mind. So it's *I love to go a-wandering along the mountain track, and as I go I love to sing, my knapsack on my back* that causes a bear to lift its head, look long in my direction, then with unmistakable dignity stroll off on a course diagonally away from mine, as if that were the way the bear wanted to go all along. *Valderi, valdera, valdera, valdi hahaha haha valderi, valdera, my knapsack on my back.*

I have decided this is not a bad way to walk in the woods—with this kind of attention, with this singing.

If my father had lived long enough to join us here in the wilderness, he would be on his knees, poking a stick into a bear pile, sorting out a beetle's wing or identifying the genus and species of the undigested plants. My father was the person who taught me to pay attention. He was a naturalist for the Rocky River parks when my sisters and I were small. His job was to lead field trips on Sunday mornings along the river or into the beech-maple forest under the approach path for Cleveland Hopkins airport. He knew every birdcall; he knew the name of every plant, in Latin and in English. He knew why only female mosquitoes bite. He knew why stinging nettles sting. It didn't matter that airliners regularly roared over the tops of the trees or that teenagers washed their Thunderbirds in the river upstream. For the people on the field trip, the morning was a great satisfaction. They found what they were looking for, which, as I think about it now, must have been an intimacy with the everyday marvelous, the miracle you can cup in your hand.

If my mother were here walking the bear trails, she would be organizing us to sing rounds. "You start, you're second, you come in third," she would say. If we are going to sing to alert the bears to our presence, we will give them a rousing chorus, not insult them with a single line of song. It is harmony, after all, that we seek. Harmony that will save us. *All things shall perish from under the sky. Music alone shall live, music alone shall live.* If my mother were here, we would march across the mudflat, keeping one eye on the bear grazing in the grasses on the far side of the cove, making harmony the way a round makes harmony, weaving the song like grass is woven in bear droppings in the spring, pausing to hold an especially beautiful chord as a gift to the bears. My mother taught us dozens of rounds, and so we learned to listen, to be attentive to the music of others, to tune ourselves to their chords, to pace ourselves to their rhythms. We learned that weaving songs is one of the most beautiful things we can do, and we can't do it by ourselves.

During World War II, my mother went to Washington to help rehabilitate soldiers. She was the nurse who led the singing therapies. I can imagine all the grievously wounded soldiers, pale on white sheets, singing *I love to go a-wandering along the mountain track*. And maybe that was healing. Maybe that was exactly what they needed, to sing in harmony from all the beds in the ward, to sing in harmony from the place of their lonely grief, the song of the Happy Wanderer.

Can it save our lives, this joyous attention?

In a literal sense, it might someday. Pay attention, and you might see a bear before it sees you. Then you can stop, group up with your friends, and find another way to go, even if it means crashing through the devil's club that stings the backs of your hands. Sing, and the bear will hear you coming even if the creek is in her ears. Then, maybe she will remember she has an appointment in another place.

So joyous attention might save someone in that small way. But there's much more to it than that. Thomas Berry wrote,

> We are most ourselves when we are most intimate with the rivers and mountains and woodlands, with the sun and the moon and the stars in the heavens, when we are most intimate with the air we breathe, the Earth that supports us, . . . with the meadows in bloom. . . . However we think of eternity, it can only be an aspect of the present.

Some cultures might find intimacy in ritual or prayer, in dance and art, in food or drink. But it seems to me that in our very peculiar Western culture, we find intimacy through understanding. Afraid of mystery, we want to know about things, how they work, how they're put together, what their names are and how they are related. Knowledge delights us. The more we learn, the more clearly we understand the density of the webs that connect one thing to another, and humankind to it all.

So there is another way that attentiveness might save us—all of us and the Earth's plants and animals. The great conservationist Aldo Leopold says that a person can't love what she doesn't know; that may be true. It doesn't follow that if she knows something well, she will love it well, but at least she will have met a necessary condition for love, which is a door open to the wonder of the world around her and her place within it. If attentiveness can lead to wonder, and wonder can lead to love, and love can lead to protective action, then maybe being aware of the beautiful complexity of lives on Earth is at least a first step toward saving the great systems our lives depend on.

Each of us is so much more than we think we are—these hands, this body, these sorrows and hopes. We are air exhaled by hemlocks, we are water plowed by whales, we are matter born in stars, we are children of deep time. How much can we improve the human prospect if we pause to notice, and to celebrate, the full extension of our being into time and the universe?

On the trail not far from our cabin, there is a place where bears measure themselves against the Earth. What you see there looks like four indentations in the trail—right paw, left paw, right—as if a Paul Bunyan bear had hammered his paw prints into the dirt as he strode by. We are told that this is exactly what it is. Year after year, the bear scuffs his paw-prints in the soft soil, placing his paws in exactly the same places each time. With depressions in the soil, he makes a record of the length of his stride and the weight of his massive body. Young bears coming by can measure themselves against the stride of the biggest and strongest, and I can imagine this, because I have seen children run and hop behind their fathers, trying to put their boots in the wide boot prints, knowing then how small they really are.

Our friends and I tried this, leaning down to put our hands in the forepaw prints, stretching our legs to reach the prints of the hind paws, and then lumbering along, butts in the air, noses to the ground, like the world's most inadequate bears. Finally, we tipped onto the ground in our yellow raincoats, laughing and afraid, humbled by the paw prints of a bear. This might be a good place for a human to be, this close to the Earth—smelling, as the bear smells, the sweetness of skunk cabbage, the seaweed-salt of intertidal flats, the melted-snow sharpness of air off the mountain; knowing, the way the bear knows, the rolling of the seasons; living, the way the bear lives, the sharply responsive life, layered with danger and possibility; glad for whatever taste of eternity is open to us, which is not to live forever but to live in grateful attention to each moment of our allotted time.

So now I'm tempted to make the argument for the import of joyous attention in another, by now familiar, way:

PREMISE 1. It's not just the sun in winter, the salmon sky that lights the snow, or the blue rivers through glacial ice. It's the small things too—the kinglet's golden crown, the lacy skeletons of decaying leaves, the hidden snakes, and the way all these relate to one another in intricate, infinite patterns. The timeless unfurling of the universe, or the fecundity of God, or an unknown mystery, or all of these together have brought the Earth to a complexity that is worthy in itself and significant in its connection to planetary, and thus human, flourishing.

PREMISE 2. What is of great worth and significance ought to be fully attended to and thus appreciated—that is, recognized as worthy and significant. Appreciate: *ad* ("to") + *pretium* ("price"). *To recognize the full worth of. To recognize the full significance of.* The opposite of *appreciate* is *disdain,* from *des* ("do the opposite of") and *deignier* ("treat as worthy"). Disdain feeds reckless disregard for the fate of what gets in the way of human purposes. Disregard allows destruction. And when the destruction is done disparagingly and in exchange for something of far lesser value, words can hardly express the appropriate moral outrage.

CONCLUSION. Some people say that we live in two worlds—the world of what is and the world of what ought to be. But to be fully attentive closes the gap between what is and what ought to be. If this is the way the world is—beautiful, astonishing, wondrous, awe-inspiring—then this is how we ought to act in that world: with attentiveness to its worth, with deep caring and fierce protectiveness, and with a full sense of our obligation to the future, that this treasure shall remain.

THE ARGUMENT GOES directly to the consequences of the disregard or disdain for the Earth, the casual ease with which corporations poison a desert spring, the casual ease with which we neighbors allow it. How easy it is to cut a tree in which nuthatches nest. How easy it is to plow a native grassland or build green-lawned houses on the feeding grounds of the meadowlark. How easy it is to build a coal dock on the spawning grounds of the salmon. If any of the agents of destruction were fully attentive to, appreciative of, the lives recklessly destroyed, would it be harder? Would it be at some point impossible? Could it become unthinkable? The industrial growth economy is empowered by the failure to pay attention to the worth of what will be consumed or destroyed by the smoking engines of progress. Is this the death culture's cardinal sin—not even to notice?

This might have been the point of the environmental impact assessment requirements—to slow the process down enough that people would notice what might be destroyed. When we were graduate students, Frank and I camped on Rabbit Ears Pass in Colorado, setting out live traps for little animals in the evening and releasing them in the morning, estimating how many would be displaced or killed by a planned water diversion canal. Perhaps that model still remains in many places. But now, it is possible for scientists to work in what I call the growing Industry of Overlooking—consulting firms where they are well paid to fail to notice (or to underappreciate) what might get in the way of the bulldozers. Of course, no one will ever know if the data were conscientiously gathered or honestly and fully reported; this research is proprietary—owned by the industry—and so is secret.

As I think about attentiveness, I've been rereading Rachel Carson's beautiful book *A Sense of Wonder,* which she wrote as an article for *Woman's Home Companion.* Most people know Carson as the author of *Silent Spring,* an argument for action to stop the pesticides that were decimating the birds. Fewer know her work that establishes the first premise for that argument—that the natural world is an entity of intricate and unfathomable worth. In books like *The Sea around Us* and *Under the Sea-Wind,* she *called attention* to the significance of the lives and patterns of life in the sea. Here she is watching a small ghost crab in the pit he has dug for himself in the sand:

> The little crab alone with the sea became a symbol that stood for life itself—for the delicate, destructive, yet incredibly vital force that somehow holds its place amid the harsh realities of the inorganic world. . . . Underlying the beauty of the spectacle there is meaning and significance. It is the elusiveness of that meaning that haunts us, that sends us again and again into the natural world where the key to the riddle is hidden.

Rachel's words, and her life, teach that attentiveness to the natural world is a source of strength, healing, and renewal. "Those who contemplate the beauty of the earth find reserves of strength that will endure as long as life lasts," she wrote. And in fact, during those terrible times when Carson struggled for the strength to continue her work even as she was dying of cancer, she found comfort at her beloved coast. There, when winter turned to spring, sanderlings stopped to feed on sandy beaches as they migrated through the ancient cycles of living and dying and living again that Carson chronicled over and over—the life cycles of the barnacles, the journeys of the eels, the necessity of death to life. "There is something infinitely healing in the repeated refrains of nature," Carson wrote, "the assurance that dawn comes after night, and spring after winter."

But the moral significance of joyous attention goes beyond its instrumental value. Philip Cafaro has made the argument that Carson's moral views fall into the tradition of virtue ethics. Following Aristotle, he defines virtues as "qualities that allow a person to fulfill his or her proper or characteristic functions and to flourish as a good of his or her kind." A virtue ethic asks after those qualities: What is a good life for a human being, and what are the personal qualities that allow a person to thrive? So also do Rachel Carson's books urge on readers a vision of what qualities might help a person thrive. Here is Rachel taking her nephew by the hand and leading him to the places that have brought her so much comfort and fulfillment, crawling on hands and knees after singing insects, showing by her example how he also might live a significant and joyful life in close relation to nature. So I think the kind of attentiveness that Rachel called "a sense of wonder" is a virtue in at least this sense, finding what it means to be fully human in a celebration of the wild world and our membership in it.

This wondering, noticing, appreciating approach to the world, she writes, is "an unfailing antidote against the boredom and disenchantment of later years, the sterile preoccupation with things that are artificial, the alienation from the sources of our strength." I think of "sterile preoccupation" and marvel that Carson could have so clearly foreseen our own time, sixty years away. The economic forces of our lives are centripetal, tending to spin us in smaller and smaller circles, creating a kind of solipsism that comes from separation from the natural world and our biocultural communities. It's not that we aren't natural creatures, it's not that we don't live always in the most intimate contact with the natural world, which seeps in our pores and rushes through our blood. It's that we lose track of that fact or deny it, and so shut ourselves off from a large part of our own humanity. We measure our successes and failures against our own mean interests, and so they grow to grotesque proportions. Self-importance and self-absorption bloat and distort our lives and our relationships.

Meanwhile, Earth turns, birds fly north or south, fish rise or sink in the currents, the moon spills light on snow or sand, and we? Do we think we turn the crank that spins Earth? Paying attention to a night of roaring waves, a face full of stars, the kick in the pants of an infinite universe, the huge unknowing, alerts us to the astonishing fact that we have any place at all in such a world. Attentive to that, we live richer, deeper lives, more fully realizing our humanity.

But Rachel has a larger point to make. The close and wondering attention she advocates offers hope also for the thriving of the beyond-human world. It is an antidote to the view that the elements of the natural world—sanderlings, shale reefs, ancient pines—are merely means to human ends, commodities to be disdained or destroyed for profit. Wondering attention reminds us of the essential worth of the world we're part of, and so it reminds us of the responsibilities that grow from that regard. It cannot "exist side by side with a lust for destruction."

THE GREAT ALPINE expanse of Denali National Park in Alaska is riven by a single road, the famous Park Road, which winds through a Pleistocene landscape of marshes and high moors. To travel this road, visitors have to pile into green park buses. However distracted they might be when they arrive at the bus stop, however obsessed by their cameras and airline schedules, when they clamber onto the bus, visitors are deeply, sometimes desperately, attentive.

"Watch carefully," the bus driver says. "There are fifty-one pairs of eyes on this bus, so collectively we have a greater chance of seeing things than any of us alone." All eyes are turned to the windows as the bus churns past the subalpine fens where moose might stand knee-deep in buttercups. All eyes are turned to the windows in the heather highlands, where bears might dig roots or caribou might look up from their grazing. Everyone is watching.

Watching. From the Old English word *waeccan,* which means "to be awake." Awake to the world outside the window, awake to the tangled plant and animal lives in the alpine's short summer, awake to the mountain light through storms. Watching is an energized form of paying attention. For the passengers, there is glad anticipation in the watching, an eagerness to see something they have seen only in books, something they have traveled halfway around the world to see, something they silently yearn for—a glimpse of a place that is functional and whole and true to itself.

The bus driver slows and stops. "Wolf!" he says. "To the right, there, in the gully at the edge of the road." Everyone—the German women, the three children from Albuquerque, the Japanese lovers in the backseat, the white-haired couples from the cruise ships—all press to the right, searching for the wolf, but searching also for a glimpse of what is wild and astonishing, searching for assurance that there is still untamed beauty and wildness in the reeling world.

"Do you see it?"

"There, half-hidden by the bush." A child sees him first. "It's sort of gray."

Then they all see him. The wolf bites at his fine tail, then sits on his haunches and gazes past the hanging-bell meadow to the rocky moraine.

There is an art to paying attention. Humans are *born* attentive, squirming wide-eyed into the world. Nothing new escapes a child's eyes. But if, somewhere along the way toward adulthood, the child starts to take the world for granted—literally that, as a given—then the great, wild world can fade from her vision, nothing more than the background of the set on which she plays out her life. At that point, it is something very much worth doing, to come to a place entirely new and practice the art of watching.

I've written an instructional flyer for the national park buses. Here it is, instruction in paying joyous attention:

THE ART OF WATCHING

1. *Watch for shadows.* Animals have evolved to hide in the tapestry of their colors, so it may not be the animals themselves but their shadows that you see. A gray wolf can disappear in gray boulders, but she cannot hide the shadow loping upside down along the gulch beside her. Watch for the black shadow of a golden eagle drawing a topographic line across valleys and moraines, the shadow of

a caribou's neck stretched by evening sun over tundra, the trapezoidal shadow of a lynx that is himself invisible on the road, fish shapes rippling on sand.

2. *Watch for reflections.* If there are swallows swooping over the pond, you will see their reflections first. It will take some searching to find the birds themselves, hidden as they are against the sweeping limbs of spruce. If there is a beaver, you will first see its wake raising a ripple on the reflected sky.

3. *Watch with your eyes shaded from the sun.* There are some things in this world that are perfectly matched to their purposes, as the baseball cap is perfectly matched to the fly ball. Up, up, you follow the arc of the ball while the crowd cheers. In Denali, the curl of the cap keeps the sun from your eyes, as up and up you follow the mounds of mountain avens into the fellfields, across the talus, to the ice-sharp ridges, to the last green cliff, to the highest pinnacle of rock, and there they are, five white specks, Dall sheep grazing slowly through the rustling applause of the endless wind.

4. *Watch especially for things that have no name.* There is no word for the little nests of leaves and mud that mark the flood line in riparian trees. Or the broken-heart shape of moose tracks in mud. Or the clapping dance of a child chasing a mosquito. There is no name for the moment on a summer night when dusk becomes indistinguishable from dawn, that blending of lavender into rose. There is no name for the unity of arctic ground squirrel and red fox. Words, the relentless nouns of the English language, divide the world into sharp-edged and familiar things. If there were no

words at all, would it be easier to understand that all things blend without boundaries into one beautiful whole—for which there is no name?

5. *Watch with your eyes closed.* While your eyes are resting, the other senses come to life. With your eyes closed as you doze after lunch, you can smell a passing grizzly—the stink of old garbage and outhouse—or the new growth of balsam poplars—sweet honey with lemon. With your eyes closed as you soak in the morning sun, you can hear water drip from the roof of the ice cave undercut by the stream—a children's bell choir, the random eagerness of the tinkling silver bells. With your eyes closed in the back of the bus, dusty and weary from the day, you can taste mountains pulverized by time. And what you do feel with your eyes closed? Maybe a thunderstorm receding, that deep, clean calm.

6. *Carefully watch for what you cannot see.* Some things cannot be seen because they are hidden: a falling fountain of northern lights, burnt out in the blaze of the sun. And some things cannot be seen because they are gone: the pregnant alpha female of the Grant Creek wolf pack, killed by a trapper at the boundary of the park. Some things cannot be seen because they are too fast: a harrier stooping on a vole. And some things cannot be seen because they are too slow: a mountain rising where North America scrapes under the bed of the sea.

7. *Watch for motion. Watch for stillness.* Like dragonflies, the first thing a human sees is motion. So watch for movement where you might expect stillness, and watch for stillness where you might

expect movement. Often, the only difference between a bear and a boulder at a distance is that the bear will lift her head to make sure her cub has not strayed. Bears swing their giant heads through the wildflowers or lope along the riverbank. But when the willow withes are bent to the wind, the bears will hold their ground.

8. *Watch for sameness and watch for difference.* Compare the color on the shoulders of a sow grizzly in sunshine with the velvet on the antlers of a caribou bedded down on the beach, or the belly of an arctic warbler, or willow catkins backlit at dusk. How is it that light, usually so imaginative, returns again and again to this buttery glow? Contrast the growth pattern of aspen trees, the dusty yellow trunk dividing into branches into twigs into leaders and leaves, with the growth pattern of streams. Tiny rills in the shadows of snowfields join streams that join waterfalls that gather in rivers that flow through gravel plains to the one great sea. Why do trees endlessly divide but rivers eternally gather?

9. *Watch with your eyes in constant motion.* Scan the landscape restlessly, the closest thickets to the far peaks. Constant motion is the trick of the veteran watcher. The highest resolution of the human eye is not in the center but slightly off to the side. So trust the flickering glance more than the lingering stare. And don't stop there. Look toward the future, when the green leaves of poplars will float like yellow boats down the river as the first snow mounds on midstream rocks. And then look into the past. The purple hanging valleys and green tundra plains speak of the Pleistocene epoch, when woolly mammoths wrapped their trunks around sedge clumps and stuffed them between their massive teeth.

10. *Watch to appreciate the world.* What is this astonishing and deeply worthy world, where northern anemones turn their faces to follow the sun, where a bear hums as she nurses her cub, where sunlight plays ice like a xylophone, where bees crawl down the throats of purple bells? A good day of watching, this glad and grateful attention—a scour of wind-driven rain, the surprise of a white mountain massif, the tundra that flowers to the edge of the known world—these remind watchers that there are interests that are not our own, there is beauty we did not create, there are depths of mystery we cannot fathom, there are wonders beyond words. This understanding may be what we are really watching for through rain running down the windows of the bus.

an old worldview, a new worldview

I'M BACK IN my Philosophy Department office, thinking about the warming Earth, the calving ice, the melting permafrost, the souring seas. How did our culture lose its way? A culture's actions, our sense of what is usual and proper, is shaped by a worldview. Some people call it a *cosmology*, a description of the cosmos that structures human lives. It's an ethos, a set of guiding beliefs that people swim in, often no more aware of them than a fish is aware of water. A worldview is a culture's answers to three philosophical questions. The questions have appeared before in this book and will again, because they structure all understanding: *What is the world? What is the place of humans in the world? How then shall we live?* Throughout history, cultures have come up with answers to those questions and then lived in those answers, in great planetary experiments, to see if they got it right. And so cultures rose and fell, converged and collided, turned, and started over.

In our time, we are at the end of a centuries-long, global-scale experiment that is testing the worldview that undergirds the industrial growth society. This is the hypothesis that humans are fundamentally different from and independent of the rest of the world, which was created solely to serve human ends. And not just different: the human species is superior to all others, the king of the mountain, high enough to reach up and touch God. Let's call this the Worldview of Separation.

Many thought leaders have traced the origins of that worldview, finding them, variously, in Greek atomism; in Christianity; in sixteenth- and seventeenth-century European Enlightenment philosophers (René Descartes and Francis Bacon notable among them); in scientific positivism; and in capitalist theory. The story of the worldview's origins is a fascinating story, well debated in other books. For our purposes, we can summarize the separatism worldview in a list of isms (the suffix by which a description becomes a doctrine, used as a derogatory word since 1670):

Dualism. There is a deep crack down the center of creation, dividing matter and spirit, body and mind. Because humans, alone in the universe, have mind, humans are fundamentally different from the rest of earthly stuff.

Materialism. Apart from humans, the world is a quarry of stones and insensate matter. Sometimes the matter can be organized like machines, to move and cry out, but it is mere matter nonetheless.

Anthropocentrism. Humans are the center, the purpose, and the culmination of creation. We are the reason for the universe. Possessed of both mind and body, we are superior to the nonhuman world. That world of so-called natural resources and ecosystem services was created for human use alone and derives all its value from its usefulness to humanity.

Technologism. Through increasing knowledge and technology, humans can harness the planet to their purposes, subdue its unruliness, and control its processes, thus amassing material wealth, which is the foundation of human happiness.

Individualism. Humans are essentially isolated rights-holders. We create ourselves most fully apart from, in conflict with, and in competition with one another.

Human exceptionalism. Because of their technological prowess, humans can make themselves exceptions to the rules that govern the lower forms of life.

Anthropocentric consequentialism. An act is judged to be right or wrong depending on its consequences. Those acts that most advance human happiness are right; the others are wrong.

Put all that together with an economic system that glorifies greed, that rewards people for shortsighted and destructive acts, and that asks us to define ourselves as consumers rather than moral agents—I'm talking about capitalism and all its invidious powers. Add powerful, amoral corporations, legal "persons" who are created without a conscience. Add faith in human progress, an inexorable march toward more and better. Now you've got the imagined world that we've superimposed on the steaming rocks. This is the worldview that our culture is testing. Can a culture flourish when it acts on the assumptions of that worldview?

We humans are the subjects of this experiment, as is every other being on land and in the sea. There are no informed consent forms.

There are no controls, as many have pointed out. There is no going back. And now the results of the experiment are starting to come in, driven by flood, storm, drought, and starvation. It's definitely starting to look like the Worldview of Separation didn't get it right. Judging by its consequences in the real world, we got it profoundly, disastrously wrong. Evidence is that a global culture built on a dualistic worldview, a philosophy of separation, will end up destroying its habitat, undermining the physical systems that support it. Derrick Jensen puts it baldly: "The first thing we can do is realize that this culture isn't and can never be sustainable, and then recognize that unless it's stopped it will kill everything on the planet."

IT DOESN'T HAPPEN often, but every few centuries or so, an entire culture goes through what philosopher Thomas Kuhn, in *The Structure of Scientific Revolutions,* calls a paradigm shift. A culture is swimming along in the waters of its worldview, but gradually the worldview doesn't seem to fit the world as well as it once did. Things become unsettled. The old ideas don't give us what we anticipated. *Turning and turning in the widening gyre / The falcon cannot hear the falconer; / Things fall apart; the centre cannot hold.* Then a crisis occurs; an undeniably unsuitable fact presents itself—for instance, the discovery that the Earth is not the center around which the cosmos revolves, or the discovery of an unknown world, or the emergence of a new prophet or technology. In our time, the "inconvenient truths" are mass extinctions, crop failures, climate chaos, collapsing food chains—we all know the list. Then the world turns upside down, the world*view* turns upside down, and revolutionary change in the foundational ideas takes place, replacing the old worldview with a new one.

It's a profoundly disturbing time when an entire culture loses its footing, as you and I know, because we are living in a time of paradigm shift. The old guiding beliefs have given way under the accumulated weight of new ecological understanding and astonishing levels of destruction, revealing themselves as metaphysically, ecologically, pragmatically impossible. Now, the search is on for a new way. Joanna Macy calls it the Great Turning:

> The Great Turning is a name for the essential adventure of our time: the shift from the industrial growth society to a life-sustaining civilization. . . . A revolution is under way because people are realizing that our needs can be met without destroying our world. . . . Future generations, if there is a livable world for them, will look back at the epochal transition we are making to a life-sustaining society. And they may well call this the time of the Great Turning. It is happening now.
>
> Whether or not it is recognized by corporate-controlled media, the Great Turning is a reality. Although we cannot know yet if it will take hold in time for humans and other complex life forms to survive, we can know that it is under way. And it is gaining momentum, through the actions of countless individuals and groups around the world. To see this as the larger context of our lives clears our vision and summons our courage.

Geologian Thomas Berry calls it the Great Work: "The Great Work before us [is] the task of moving modern industrial civilization from its present devastating influence on the Earth to a more benign mode of presence." And again, "Our own special role, which we will hand on to our children, is that of managing the arduous transition from the terminal Cenozoic to the emerging Ecozoic era, the period when humans will be present to the planet as participating members of the comprehensive Earth community."

The *terminal Cenozoic*. How extraordinary to be in this time and place, when immortal beliefs about who we are in the world are revealing themselves to be not only precariously mortal but lethal as well. How extraordinary to be in the generation assigned this task of world transformation, even as the ground we stand on is trembling with a global paradigm shift.

Several years ago, I invited two dozen people to come together in Oregon's ancient forest to problem-solve about the failure of the usual ways of thinking to deal with climate change. Metaphorically, the question we faced together was, *In the "perfect moral storm" of climate change, what do you do when the life rafts are burning?* One of the people I invited was a cherished friend, the Alaskan singer-songwriter Libby Roderick. Libby desperately wanted to bring a new song to this gathering. But her songs were about hope, and she found she had nothing left to offer. For two months, she wrestled with the challenge, knowing that she does not simply write songs; they come to her slowly, from a deep place. The night before she was to leave for the gathering, she sat at the piano and said to the universe, "If you want something from me, you have to tell me." Weeping, she began to sing the questions. It's hard to ask terrible questions, and harder still to wait quietly for the answer. Late that night, the answer that came to her was "Turn." Here are the words to the song she brought as a gift to our questioning:

WHAT DO YOU DO WHEN THE LIFE RAFTS ARE BURNING?

What do you do when the life rafts are burning and your
babies are inside the boat?
Where do you turn when the waters are churning and
you're desperate to learn how to float?
How do you pray when your prayers go unanswered and
each crier feels so alone?
What do you do when the life boats are burning?
You turn, turn, turn.

Turn to each other, reach for each other, take one another
by the hand.
We'll form a life raft, a human life raft, and we will swim
toward land.
If we make it, we start over. If we don't, we go down
together,
One for all, all forever, turn.

What do you do when the iceberg is looming and the ship
isn't turning away?
How can you be heard when the warning bell's ringing and
the band just continues to play?
Where do you turn when the engines catch fire and the life
rafts are starting to burn?
Turn, turn, turn.

Turn to each other, reach for each other, take one another
by the hand.
We'll form a life raft, a human life raft, and we will swim
toward land.

If we make it, we start over. If we don't, we go down
together,
One for all, all forever, turn.
One for all, all forever. Turn.
One for all, all forever.

If we make it, we start over. The results of our cosmic experiment in separation are asking our industrial growth society to rewrite its guiding assumptions about what the biocultural world *is* (What is the world? What is the place of humans in the world?). At the same time, the devastating failure of the current worldview asks society to start over with our ideals about what the world *might become* (How then shall we live?).

START HERE: LET'S put to rest once and for all the separatist notion that we humans are apart from, different from, masters of, the great systems of the Earth. Such arrogant and preposterous claims have mapped out a miserable, lonely world. For three, four hundred years, we Children of Separation have believed that we lived alone and damp-eyed in a world of nothing but stone and dumb brutes, the only spirit in a universe of matter and mechanical animal-clocks, the only shining eyes in a universe stripped of mystery, exposed to human understanding and control, reduced to human convenience. Lonely kings of the stony mountain, we industriously transmogrified nature into wealth and human biomass.

That's over. Now ecological science, indigenous wisdom, and almost all the religions of the world call us to recognize our kinships within the interconnected planet. How complicated and layered and open-ended this kinship of humans with all of natural creation actually is, this beautiful, bewildering family.

First, there is the kinship of common substance. Like a sea slug or a horseneck clam, I am carbon atoms spun through time, arranged and rearranged in patterns. Break my pattern down to atoms, and I can scarcely be distinguished from the stars. Second, there is the kinship of common origins. The gooseflesh that prickles my skin is what's left of the contraction that bristles the fur of a frightened bear and fluffs a bird in February. Third, there is the kinship of interdependence. Consider the sweet-rotting hemlocks that create vanillin, which nourishes the mycelia, which produce fruiting bodies, which feed the flying squirrels, which nourish the spotted owls, which entrance a middle-aged woman, who warms the rotting tree with her fleece bottom. And fourth, the kinship of a common fate. We, all of us—blue-green algae, galaxies, bear grass, philosophers, and clams—will someday dissipate into vibrating motes. In the end, all of natural creation is only sound and silence moving through space and time, like music.

The same arguments get trotted out, over and over again, in a dogged effort to preserve a place on the pedestal for humans alone. Here's the old favorite: It would seem that humans are apart from and above the rest of natural creation, for the Bible says that God created man in his own image, several days after he created the birds of the air and the fish of the sea, and gave man dominion over all the creatures that walk and fly and swim.

But I would say it is only arrogance that makes humans think that because God made us last he must have made us best. Why couldn't people as readily believe that by the time God got around to making people he had run out of ideas, so he did a little recycling?

The same mechanical systems that propel giant squid through the seas move blood through our hearts, and our cells use the same notation that directs the growth of red rock crabs. We breathe the same oxygen as the fish of the sea and the birds of the air, and exhale the same carbon dioxide from our lungs. To make human minds, our exalted minds, God blew into the brain of a lizard. Our great temples have the proportions of a snail.

If humans truly are made in the image of God, then God too must be made of the stuff of lizards, or lizards are made of the stuff of God. We should rejoice in every squid and sculpin—the progenitors of what is divine in us.

A different argument comes from Descartes. Humans have minds, or consciousness, he wrote, "the thinking substance." But plants and animals do not. So humans are apart from and superior to plants and animals.

The fact is, I don't know for sure what animals are thinking, but neither did Descartes, and that seems like a good reason not to rush to judgment about what's on an animal's mind. How suspiciously convenient it is to believe that humans have the monopoly of the universe on mind. If people are going to imprison dolphins and grind the gallbladders of bears into fortifying elixirs, if they are going to scrape the bottom of the ocean bare and squeeze the hindquarters of black-tailed deer into patties, if they are going to reduce owl nesting sites to toilet paper and convince themselves that this is not a problem, then they will need to believe that humans have minds but other animals do not. But this is a matter of convenience, not truth.

I'M WRITING THIS evening in my Alaska cabin. When I step out onto the porch and listen, this is what I hear: Seawater drips from sea palms. Rain ticks through hemlocks. A man and his son murmur in lantern light. A winter wren startles from sleep. A kelp crab shuffles under sugar kelp. A boat creaks in the pull of the tide. It's all one symphony—all the small sounds sounding in the darkness, coming together to create one beautiful night. That's when I know that whatever harm we do to any part of the music, we do to the harmony of the whole. The discordance we create will be in our lives and the lives of our children.

OKAY, NOW LET'S see what dents I can make in the notion that human happiness is the only measure of right and wrong.

I submit that the times call out for a different answer to the foundational question, *How then shall I live?* The usual ways of thinking have allowed us to drift into this terrible storm. Worse, ethics-as-usual, anthropocentric consequentialism may have driven us into the teeth of that very storm.

But what other ways are there to think about whether an act is right or wrong? Several, as it turns out. It's helpful to remember that an act doesn't take place at a single point in time. A given act begins in a person's *character,* the habits of mind and heart that define him as an individual. His character gives rise to an *intention* to act, and that intention results in the *act* itself. The *consequences* of the act stretch infinitely into the future—consequences for humankind and consequences for the biophysical world.

So there is a whole string of factors to think about, beyond the human consequences. Those factors yield at least three different moral theories:

Virtue theory focuses on the character of the actor. An act is right if it springs from a virtuous person with good intentions.

Duty-based theory focuses on the nature of the act itself. An act is right if it conforms to duties, which may be defined by God, by justice, by social contract, by promises, or by something else.

Consequentialist theory focuses on the results of the act. An act is right if it creates the greatest good for the greatest number. A biocentric, Earth-centered consequentialism pays attention to the good of all beings and the strength of their connections. An anthropocentric, people-centered consequentialism pays attention only to the good of human beings.

The problem is that the industrial growth culture has generally ignored considerations of virtue or duty and crowded itself into the narrowest possible corner of moral discourse—the corner where policies are weighed by their projected consequences in a giant risk–benefit analysis, a computation of human benefits and burdens that does not take the rest of the world into account.

But this isn't working anymore. How does one weigh the benefits and burdens of a decision when the benefits are here and now and the burdens are far away in place and time? Can we discount, and to what degree, the costs to the future? Are we to weigh costs to this generation against costs to the next, the next one hundred, or an infinitude of generations? How do we calculate the benefits and burdens when we have only the slightest idea of how our decisions will play out in the confused seas of the future? These are questions that consequentialism itself cannot answer. And worse, if we can't see that our individual decisions will make a difference, consequentialist moral theory can't ask anything of us at all.

But it's the exclusive focus on the well-being of humans that sinks this ship. Disregard for the well-being of any part of the world but the human has led us to sail merrily toward ruin. So even on its own terms—a right theory is one that will increase human well-being—anthropocentric consequentialism fails. We once might have thought that we had no obligations except to ourselves, as individuals or as a species. But we now understand that humans are part of intricate, delicately balanced systems of living and dying that have created a richness of life greater than the world has ever seen. As humans, we are created and defined by our relation to those great systems; we find our greatest flourishing in cultural and ecological community. We should probably not be surprised that moral theories devised to fit a separatist view of the world are not serving us as well in the holistic world that we are beginning to understand.

Conservation scientist Aldo Leopold defined the first premise of any ethic as this: "that the individual is a member of a community of interdependent parts." If we are on the verge of understanding that deeply, then we are on the verge of a new ethic and we are on the verge of inventing new forms of human goodness. If we can do this, we might be launching a new experiment, testing whether human civilizations can thrive in concert—take that phrase seriously, *in concert*—with the rest of creation.

If we make it, we start over. Okay, let's start over. And let's begin with ethics. If the world is an inanimate bundle of rocks and squeaking things that has no value in itself, and if human beings are the lords and masters of this world, then how should we live? We should use it to our own ends, without limit or conscience. Fine. But if we understand the world differently, as a place of embedded systems that are interdependent, finite, resilient, and very beautiful, and if we are members of those communities, fully participating in their thriving and fully subject to their rules, we will act very differently.

What is needed is an Earth-centered ethic that is strong and supple enough to support the hard decisions we will have to make. Who do we aspire to be? Who do we need to be? What are the emerging forms of human goodness? What are our obligations of gratitude and respect? What do we hope for? Can we imagine an ethic that spans the whole universe of moral possibility, not just a single corner? Can we imagine an ethic that is inspired by, even patterned after, the Earth itself?

an ethic of the earth

HOW COULD ONE begin to write a new ethic that is aligned with the way the world works rather than in defiance or denial of it? "Nothing so important as an ethic is ever written," Aldo Leopold wrote in *A Sand County Almanac*. "[An ethic evolves] in the minds of a thinking community." If we need a new ethic, my colleagues and I decided, then we need to convene that thinking community. And if it's evolution that needs to happen, we need to crank up the rate at which evolution occurs. We invited twenty-two of the nation's creative moral thinkers—this time including not just philosophers but also activists, theologians, writers, poets, and scientists. Our ambition was to jump-start the evolution of a new ethic that responds more powerfully to the environmental emergencies we face and so has the power to carry us to a lasting, just, and joyous biocultural Earth community.

So here they all came, into the mossy shade of five-hundred-year-old trees growing in the valleys of the Blue River at the H.J. Andrews Research Forest. It was raining, of course, but sometimes the sun speared through the trees, making the whole world steam. That first night, the rain let up and we built a fire in a forest opening. Passing a weak flashlight, we did our best to read aloud the poems and passages we had brought to give shape to our work, including this one from Henry Beston's *Outermost House:*

> The economy of Nature, its checks and balances, its measurements of competing life—all . . . has an ethic of its own. . . . Whatever attitude to human existence you fashion for yourself, know that it is valid only if it be the shadow of an attitude to Nature. . . . The ancient values of dignity, beauty, and poetry which sustain it are of Nature's inspiration; they are born of the mystery and beauty of the world. Do no dishonor to the earth lest you dishonor the spirit of man. Hold your hands out over the earth as over a flame. To all who love her, who open to her the doors of their veins, she gives of her strength, sustaining them with her own measureless tremor of dark life. Touch the earth, love the earth, honor the earth, her plains, her valleys, her hills, and her seas; rest your spirit in her solitary places. For the gifts of life are the earth's and they are given to all, and they are the songs of birds at daybreak, Orion and the Bear, and dawn seen over ocean from the beach.

For three days, we met in pairs and small groups and in our entirety, exchanging ideas online and on slips of paper, brainstorming as we ate breakfast or walked the trails, writing at sheltered picnic tables, revising far into the nights, which were nights of owl calls, tapping keys, tapping rain. For three days, we gathered and sifted the wisdom that people brought with them from Native American traditions, Latin American culture, ecological science, philosophical courage, poetry, science fiction, religions pagan and Christian and Asian, and the radical imagination of the streets. Here is what we wrote:

THE BLUE RIVER DECLARATION

A truly adaptive civilization will align its ethics with the ways of the Earth. A civilization that ignores the deep constraints of its world will find itself in exactly the situation we face now, on the threshold of making the planet inhospitable to humankind and other species. The questions of our time are thus: What is our best current understanding of the nature of the world? What does that understanding tell us about how we might create a concordance between ecological and moral principles, and thus imagine an ethic that is of, rather than against, the Earth?

What is the world?

In our time, science, religious traditions, Earth's many cultures, and artistic insights are all converging on a shared understanding of the nature of the world: The Earth is our home. It will always be our only source of shelter, sustenance, and inspiration. There is no other place for us to go.

The Earth is part of the creative unfolding of the universe. From the raw materials of the stars, life sprang forth and radiated into species after species, including human beings. The human species is richly varied, with a multitude of persons, cultures, and histories. We humans are kin to one another and to all the other beings on the planet; we share common ancestors and common substance, and we will share a common fate. Like humans, other beings are not merely commodities or service-providers but have their own intelligence, agency, and urging toward life.

We live in a world of nested systems. Living things are created and shaped by their relationships to others and to the environment.

No one is merely an isolated ego in a bag of skin but something more resembling a note in a multidimensional symphony.

The world is dynamic at every scale. By processes that are probabilistic and often unpredictable, the world unfolds into emergent states of being. Our time of song and suffering is one such point in time. The life systems of the world can be resilient, having the ability to endure through change. But changes create cascades of new events. When small changes build up and cross thresholds, sudden large transformations can occur. Thus the world in its present form—the world we love and inhabit—is contingent. It may be, or it may cease to be. If the Earth changes in ways that undermine our lives, there is nothing we can do to change it back.

The Earth is finite in its resources and capacities. All its inhabitants live within its limitations and by its rules. And although life on Earth is resilient and robust, rapid irreversible changes and mass extinction events have occurred in the past. As a result of ignoring the Earth's boundaries, we are causing a new transformation of the Earth and the sixth mass extinction of its beings.

Our knowledge of the Earth will always be incomplete. But we know that the world is beautiful. Its life forms, unique in the universe, are wonderful in their grandeur and detail. It follows that the world is worthy of reverence, awe, and care.

Who are we humans?

We humans have become who we are through a long process of biological and cultural evolution. As do many other social species, we possess a complex and sometimes contradictory set of possibilities. We are competitive and cooperative, callous and empathetic, destructive and healing, intuitive and rational. Moreover, we are creatures of consciousness, emotion, and imagination, beings through whom the universe has evolved the capacity to celebrate its own beauty and explore its own meaning in the languages of science and story.

We are dependent on the sun and the Earth for everything. Without warmth, air, water, and fellow beings, we would quickly die. At the same time, we are cocreators of the Earth as we know it, shaping with our decisions the future of the places we inhabit, even as our relation to those places shapes us. In this way, we are members of a community of interdependent parts.

Humans are beings who search for meaning. Our beliefs about the origins of the cosmos influence the way we relate to each other, to other living things, and to the habitats we both depend upon and constitute. Sometimes we experience wonder and awe at the mysteries of the universe and fall silent in reverence. Yet, as we strive to comprehend the world, we often divide it into hierarchies of value—pure/impure, spiritual/material, human/subhuman. Although we often exclude and exploit those we judge less valuable than ourselves, we yearn for belonging.

We are born to care. From the first moments of our lives, we seek connection. We deeply value loving and being loved. We find comfort in close connection to other people, other species, and to the wild world itself.

We are adaptable and resilient. Our imagination gives us the ability to envision alternative futures and to adapt our behaviors toward their achievement. When we are at our best, we develop cultural systems in which we, other living beings, and ecosystems can flourish.

We are moral beings. We have the capacity to reason about what is better and worse, just and unjust, worthy and demeaning, and we have the capacity to act in ways that are better, more just, more worthy, more beautiful.

Because we are these things, we can change. Because we are these things, change will be difficult.

How then shall we live?

Humanity is called to imagine an ethic that not only acknowledges but emulates the ways by which life thrives on Earth. How do we act when we truly understand that we live in complete dependence on an Earth that is interconnected, interdependent, finite, and resilient?

Given that life on Earth is interconnected, we are called to affirm that all flourishing is mutual and that damage to the part entails damage to the whole. Accordingly, our virtues are cooperation, respect, prudence, foresight, and justice. We have the responsibility to honor our obligations to future generations of all beings and take their interests into account when we reflect on the consequences of our actions. To discount the future, to take all we need for our own well-being and leave nothing for others, is unthinkable. We should take only what the Earth offers and leave as much and as good as we take. To live by a principle of reciprocity, giving as we receive, re-creates the richness of life, even as we partake of it. Then our harvests are respectful and thoughtful. We learn to listen, which means that we learn to value congeniality, patience, fairness, and moral courage, which creates the possibility of heroism in the face of disagreement and discord. Moreover, the new ethic calls us to remedy destructive distributions of wealth and power. It is wrong when some are made to bear the risks of the recklessness of others, or assume the burden of others' privilege, or pay with their health and hopes the real costs of destructive practices.

Given that humanity is inescapably dependent on the Earth for gifts both material and spiritual, we are called to be grateful and humble. To be grateful is to express joy for the fertility of the Earth, to be attentive to its gifts, to celebrate its bounty, and to accept responsibility for its care. Humility is based on an understanding of our own roots in the soil; we recognize the danger we face and the damage we do when we get that wrong. So we are well advised to be humble about our claims to knowledge; and with art and heart and science, to strive for continuous learning that is open to evidence from all ways of knowing and from the Earth itself. The generosity of the Earth models generosity in our relations with others and calls for collective outrage when we fail in that duty. A new ethic calls us to defend and nurture the regenerative potential of the Earth, to return Earth's generosity with our own healing gifts of mind, body, emotion, and spirit. We can find joy and justice in sustaining lives that sustain our own.

Given that the Earth's resources and resilience are finite, human flourishing depends on embracing a new ethic of self-restraint to replace a destructive ethos of excess. Greed is not a virtue; rather, the endless and pointless accumulation of wealth is a social pathology and a terrible mistake, with destructive social, spiritual, and ecological consequences. Limitless economic growth as a measure of human well-being is inconsistent with the continuity of life on Earth. It should be replaced by an economics of regeneration. Simple lifestyles that include thriftiness, beauty, community, and sharing are pathways to happiness. Celebrated virtues are generosity and resourcefulness.

Given that life on Earth is resilient, humanity can take courage in Earth's power to heal. We can find guidance in the richness of diverse cultures and ecosystems, if we honor and protect difference. Equality and justice are necessary conditions for civilizations that endure, and truth-telling has strong regenerative power. Virtues we can embody are human courage, creative imagination, and perseverance in the face of long odds. The effect of humans on the land can be healing; our obligation is to imagine into existence new ways to live that create resilient and robust habitats. If we can undo some of the damage we have done, this is the best work available to us. On the other hand, damaging the natural sources of resilience—degrading oceans, atmosphere, soil, biodiversity—is both foolhardy and an offense against the future, not worthy of us as rational and moral beings. If hope fails us, the moral abdication of despair is not an alternative. Beyond hope we can inhabit the wide moral ground of personal integrity, matching our actions to our moral convictions. Through conscientious decisions, we can refuse to be made into instruments of destruction. We can make our lives and our communities into works of art that express our deepest values.

The necessity of achieving a concordance between ecological and moral principles, and the new ethic born of this necessity, calls into question far more than we might think. It calls us to question our current capitalist economic systems, our educational systems, our food production systems, our systems of land use and ownership. It calls us to reexamine what it means to be happy and what it means to be smart. This will not be easy. But new futures are continuously being imagined and tested, resulting in new social and ecological possibilities. This questioning will release the power and beauty of the human imagination to create more collaborative economies, more mindful ways of living, more deeply felt arts, and more inclusive processes that acknowledge the ways of life of all beings. In this sheltering home, we can begin to restore both the natural world and the human spirit.

❖

I WANT TO tell you that when the twenty-three of us writers stood in a circle and read the finished declaration round-about, each of us reading a sentence in a spiral of rising conviction, until the sentences were all said and the room was quiet, I felt a swell of hope that was foreign to me. Here is just one change: to understand that we are not lonely lords but rather kin in a family of living things, aware in a world of awarenesses, alive in a world of lives. But from that one change flows a cascading river of transformation, when we start to realize all that this new perspective offers us and asks of us. I am used to thinking of a tipping point as a bad thing, a point where the board teeters on the fulcrum and then crashes to the ground. But it's the nature of a tipping point that every descent to a hard landing sends the board flying upward too. A small change creates more change, and that creates more, and then there is no going back. That must be what the Great Turning looks like.

an ethic of the cosmos

JUST AS THERE is a seamless connection between a human being and the planet, there is a seamless connection between the planet and the cosmos. And as much as we are creatures of the Earth, we are creatures of the stars. So, just as a human-based ethic drew an artificial boundary around humans, an Earth-based ethic might be said to draw an artificial boundary at the light blue edge of the atmosphere. That said, I wonder if it makes sense to think about expanding even further the sphere of our moral universe, to embed an ethic of the Earth within an ethic of the cosmos. What might an ethic of the cosmos look like?

Let's begin that thought experiment with a memory from a cold November night some years ago. Frank and I had decided to camp in the mountains to see the Leonid meteor showers. That's when Earth would plow through dust from an exploded comet, putting on a show that astronomers predicted would be the most spectacular in a thousand years.

ONE NIGHT, OF THREE HUNDRED SIXTY-FIVE

As soon as we arrived at the meadow beside the lake, we hauled our gear through dried reeds and grasses, looking for the very best place to put our bed. It didn't have to be dry or out of the wind, but it had to have a full view of the sky. On a rise a hundred yards from the little lake, with only one scraggly pine to block our view, we spread out ground cloth, foam pads, and sleeping bags and tucked a nylon tarp around the whole sliding stack to keep off the frost. Then we hurried to collect firewood, prying it from the frozen ground.

We stood in the smoke of a small fire and drank cold beers with our gloves on. In the dusk, tundra swans flashed white as they dropped to the water. There must have been hundreds of swans, a winter storm of swans, and equal numbers of Canada geese, all sounding off—swans *doo-whoop*ing, geese clattering, an occasional *thump* from a hunter's gun.

In November, dusk is long and deceiving. The lake disappears first, then the ground, then the mountains, and soon enough the fire itself becomes just an optical illusion, all light and no heat, and that's all you can see—the trembling coals and yellow toes of your own frozen boots. We kicked out the fire and groped our way to bed.

Using only the heat sucked out of your body, it takes a long, miserable time to warm a goose-down sleeping bag. There are two strategies here. One is to wiggle and wiggle, hoping the heat your exercise generates will warm the bag. The other is to hold perfectly still, concentrating every unit of heat in the fabric closest to your body. I wiggled and Frank held still, and between the two of us, it took a long, long time to get warm. The birds were making a ruckus out on the lake, yelling and laughing as the night got colder and darker. We lay flat on our backs, watching the sky through the narrow slits between our hats and the bags pulled over our noses.

The sky filled with stars, but for the moment they seemed to be staying in place. Every time I blinked, more stars appeared. "Somebody once said," I told Frank, "that since space is infinite and there are stars throughout space, the only reason the sky isn't paved with starlight is that light from the farthest stars is still on its way. Someday, when all that light finally gets here, the night will be solid starlight."

"Except for dark matter," Frank said, but I didn't care. The sky was making good progress toward infinitely starry, and I was getting sleepy.

When I woke up next, white fog covered the lake like milk. I could still see the dark silhouettes of mountains and treetops poking out of the fog. But even as I watched, the air next to the ground around us gradually went milky, until ground fog closed over us too and we lay under a thin pool of damp white air. Our next chance to see this meteor show would be November 2099. "Better start exercising and eating broccoli," Frank said, but suddenly a meteor catapulted the width of the sky. We could see it through the fog, fuzzy gold like a Christmas tree garland unaccountably come to life. After a long time, we saw another. And then one more, but we were sure we were missing most of them. We debated without much enthusiasm: Should we get up, hike to a cold car, and search for a place where the fog had cleared? But we were finally warm, and too comfortable, and we tucked down and went back to sleep.

When Frank woke me several hours later, the night was clear and black. Bright streaks of light rushed in all directions. The meteors looked like minnows to me, the way they darted and disappeared—sun-striped minnows startling and streaking off in a night-dark mountain lake—and I was at the bottom of the lake, cold currents eddying over my face. Frank counted 1 meteor every ten seconds, 6 a minute, 360 an hour. "It's dust," he said. "Just imagine: six-billion-year-old dust burning like that." An incoming fireball dropped in hard and exploded over our heads, and we yelped—in fear or wonder, I don't know. It was cold and the edge of my sleeping bag was frosting up. Still the stars fell all around us. They fell and fell. Every time I opened my eyes, they were falling.

I woke up in a silver haze. Clouds ringed the mountainsides. As tundra swans whooped on the lake, a skein of geese slowly untangled over our heads. The soil was a cathedral of crystals growing upward, pushing little roofs of dirt. Brilliant hoarfrost covered the entire world, sparkling on our sleeping bags, our hats, each branch of every tree, every bent reed. It was as if all those twinkling white stars that fell out of the sky, all that bright six-billion-year-old interstellar dust, had fallen glittering on the branches while we were sleeping, dusted our hats, drifted into the folds of our blankets, as if fallen stars had stuck to spiderwebs and outlined every blade of grass.

Sometimes I think I see a miracle, and then I realize that it's just the everyday working of the world. Warm moist air encounters cold air and turns to shards of ice. Light catches on the crystal planes. Tundra swans fly in to feed. Geese cry out. The lake reflects the sky. That's the real miracle: that it's no miracle at all. Just Earth, sailing on in the dark.

❖

I HAVE A friend, a pianist, who can't sleep under the stars for fear she will fall into their dazzle and disappear. But I loved sinking into that black, starry night. It allowed me a very different view of my husband and myself—two little *Homo sapiens* lying on their backs in a story spinning from the deepest time. That long story of the cosmos—from the first explosion of the vanishingly small point that held all things to the swirling, heat-addled winds at the end of the Cenozoic era on Earth—is a story that has the Copernican power to change the course of history.

Here, again, I am inspired by the work of Mary Evelyn Tucker. She is a professor at both the Yale Divinity School and the Yale School of Forestry and Environmental Studies, a scholar of the great religions of Asia, a coauthor of the Earth Charter, and an Emmy-winning writer of the film *Journey of the Universe*. Standing six feet tall, with curling red hair, a towering figure in every sense of the word, she has made it her life's work to reach out to the reeling world, telling of a new cosmology, a new worldview, to replace the human-centered narrative that has taken us to this destructive end.

To put her work into context: Analysts, notably Joanna Macy, say that to turn away from the destructive course of history, we have to do three things, and all of them are essential. First, stop the damage to the planet's life-supporting systems. Second, imagine new and better ways to live on Earth. Third and most important, change the story about who we are, we humans—not the lords of all creation but lives woven into the complex interdependencies of a beautiful, unfolding cosmological system. Many people are pursuing the first two goals. But Mary Evelyn and her colleagues pursue the third, offering a cosmologist's view of the universe and its children.

What is the world? Almost fourteen billion years ago, the infinitely small point that was everything exploded, slinging out dust and light in every direction. Mary Evelyn calls it "a roaring force from one unknowable moment." Over unimaginable expanses of time, the dust organized itself, spinning galaxies, and from galaxies, stars, and from stars, planets. Properties emerged from self-organizing systems, one might say, and I stop here to wonder at how it can be that matter and energy can organize themselves into something different, something more complicated, something that can reorder itself in turn. In ways that are mysteries to me, the great creative urgency of the cosmos unfurled much as a peony unfurls, layer after layer of translucent petals—not just petals but shapes and systems of unimaginable creativity, bringing forth the next layers of complexity and becoming.

Creation continued to unfold: At some point, the sizzling Earth. At some point, rain. At some point, Life. Whether plunging to Earth in an asteroid or tangling into existence at the bottom of the sea, here is Life, simple systems with the ability to replicate themselves and with some emergent form of awareness. Emergent, increasingly complex, until in this time, our time, the Cenozoic era of multicellular lineages unfolds. Silver willows, spirochetes, tall fescue, crickets, and this astonishing human creature who is both the same as and different from the others.

Maybe the process has no end point, no goal; that seems likely to me. But that doesn't imply that it has no meaning. There can of course be meaning, even in the absence of purpose. Like music, the unfurling of the cosmos is splendid in itself, "staggering," Mary Evelyn says. "That story . . . has the possibility to give us a perspective that almost nothing else does. This is the ultimate wake-up call to Life and its complexity and its beauty. This is the ultimate wake-up call to the holiness of Life, with both its *w* and its *h* forms—both whole and holy. We don't have language to fully describe the importance of this understanding."

What are humans in relation to the universe? We are beings created by the same processes of creative evolution as the nut trees and the dragonflies, created from the same substance as the stars. In us the universe has created not just awareness but self-awareness. We are the beings through which the universe can celebrate its beauty, explore its meaning, and marvel at its story.

How then shall we live? We are also the beings who can ignore the beauty of the universe, co-opt its meaning, and marvel only at ourselves and our power. Can the story of the journey of the universe offer any guidance for how we ought to live?

It calls us, Mary Evelyn says, to align our practices with evolutionary processes, instead of to bring them to an end.

From her office at Yale, Mary Evelyn explained this to me. I was supposed to be interviewing her, but all I could do was listen. Here is part of what she said:

> The creativity of evolutionary processes and self-organizing dynamics is mirrored in our own creativity, which is birthed from these processes of billions of years. Thus our creativity is enlivened by being in resonance with a life-generating planet—the flowers, the leaves, the waving grasses, the sunsets, the wind.
>
> Humans desire, more than anything else, to be creative, and we desire to participate in the creative processes, in the future and in life—that's what having children is about. But we can be life-generating in a variety of ways—creative, participatory, oriented toward something larger than ourselves.

What is larger than ourselves that we really care about? It's life, as far as I can see.

We are on the verge of knowing how to express comprehensive gratitude, acknowledging that we are dwelling within a living system. This gives rise to a sense of resonance with life-forms that certainly earlier peoples understood, and native peoples still do. This is a new moment for our awakening to the beauty of life that is now in our hands. And because we are life-giving humans and care about our children and their children and future generations of all species, I think the universe story can sustain us and inspire us in so many ways yet to be fully discovered.

Our work is to align ourselves with evolutionary processes instead of standing in their way or derailing them. So our human role is to deepen our consciousness in resonance with the fourteen-billion-year creative event in which we find ourselves. Our challenge is to construct livable cities and to cultivate healthy foods in ways congruent with Earth's patterns. We need the variety of ecological understanding so we can align ourselves with the creative forces of the universe.

Something is changing; an era is changing. If we are shutting down the Cenozoic era . . . the great work is to imagine how the new era can unfold. Our work in the world is not just a stopgap to extinction. . . . We are part of the Great Work, as Thomas Berry would say, of laying the foundation of a new cultural era.

Mary Evelyn's notion of aligning our work with the creative forces of the universe offers new approaches to agriculture, urban planning, peacemaking, and educating. A farm, for example, might be a battleground for a war against the creative forces of the universe, a war waged with plows and herbicides, insecticides and overwatering, massive amounts of nitrogen fertilizer, massive amounts of debt—agricultural industries doing to the land what the victorious Romans did to Carthage when they plowed the city and sowed it with salt. Or a farmer might decide instead to raise food in a deep and respectful alliance with the creative forces of the universe, an alliance nourished by attentiveness to the way the natural world creates grains and grapes from minerals that fell from the stars.

Align, to arrange things in a line, to go in the same direction, side by side. Might this next worldview lead us to expect that humans, as they look with stunned gratitude to the cosmic creativity that spins life from stones and sunshine, will design their work to progress with, rather than fight against, the currents of ongoing life?

LAST FALL, WHEN I returned to the edge of the lake where the meteors fell down on us, it was a strangely different place. A wildfire had stormed through the forests, a catastrophic crown fire of smoke and flames that roared around the lake, engulfed the mountains, and reduced the forests to thickets of black spars. The lake was a single blue eye, staring from an ashen, stubbled face, half-blinded by the sun. I heard no sandhill cranes or coots, just the rustle of a million bark beetles chewing tunnels in the burned spars. In that sun-glare, there were no stars to be seen.

"This is a broken world and we live with broken hearts and broken lives," the songwriting poet Leonard Cohen once stated, "but still that is no alibi for anything." The great creative processes of the universe will make something new from the charred star-matter of this lakeshore. It will be green and growing. It will catapult its seeds willy-nilly into the mud. It will rise through the water from snakelike roots and unfurl wrinkled leaves. It will shiver over gravel beds, dripping eggs. And in the revolving mornings of the Earth, there will be humans who get up from their beds and take up the work of imagining something new or recovering something old, something that will turn their culture into alignment with the ancient, beautiful creativity of the universe.

ethics and extinction

IMAGINE THIS: A giant sea scorpion lumbers through the mud at the edge of a bay, dragging its carapace on jointed legs and pushing with its oversized swimmerets. A wave of heat scorches it; it recoils. With a hissing roar of steam, a wall of burning basalt crumbles into the sea.

That sea scorpion and 96 percent of the species on Earth were wiped out in the Permian extinction 252 million years ago, in a series of events so catastrophic that it is now called the Great Dying. Molten basalt flooded from the greatest volcanic eruptions in Earth's history, releasing huge quantities of deadly methane and enough carbon dioxide to raise the temperature of the planet by ten degrees Celsius. Four percent of the species alive at that time survived the souring water and drying lakes. They provided the genetic material for all the species to come.

Or imagine this: A hadrosaur plunges her duck bill into a mass of flowering plants, oblivious to the shadow of an eight-kilometer-wide asteroid that arcs overhead. She bares her teeth and hisses at another dinosaur that is moving in to feed. A pterosaur wings out of a giant tree fern. Maybe he feels in the quivering trees the impact of the asteroid on the other side of the Earth and flies off, startled by something he has never felt before.

This time of extraterrestrial impacts and extreme volcanic activity is the end of the Cretaceous, the K-T extinction, sixty-six million years ago, when perhaps 80 percent of the species vanished, including most dinosaurs and many of the small creatures of the seas.

Awful forces of exploding rocks, boiling seas, poisonous clouds, ending forever the possibilities in the strange and wonderful bodies of a profusion of creatures: five massive extinctions in the history of planetary life, when evolutionary development started over with what was left. And now, we are told, we live in the time of the sixth great extinction, bringing the Cenozoic era to a close. How astonishing to find ourselves in this time.

"It's our generation that is witnessing the end of the era that we evolved in," Thomas Berry wrote. "My generation has done what no previous generation could do, because they lacked the technological power, and what no future generation will be able to do, because the planet will never again be so beautiful or abundant." Since 1970, 40 percent of the plants and animals—individual beings, not species—have vanished from the face of the Earth.

I remember 1970. I remember John Lennon's announcement that the Beatles were through. I remember the heart-thumping peril of Apollo 13. I remember the first Earth Day. I remember when scrub jays nested in our laurel hedge and squawked at the neighborhood cats. I remember when there was a meadow, complete with meadowlarks, where there is now an asphalt parking lot for the new Home Depot. I remember the clam flat where there is now a liquid natural gas terminal on rock fill at the edge of Newport Bay. I remember the backwater slough that is now riprap at the edge of the Willamette River. I remember the heron rookery on the headland where there is now a subdivision called the Rookery. I remember pintail ducks in our pond. All the little lives, transmogrified into money or human flesh.

But sometimes I forget to grieve. Sometimes I take it for given that this is the way of the world. I forget the call to life, the urgency to continue that is built into every plant and animal, the reaching toward life. I forget that in a different world, a better world, this level of human-caused destruction would be unthinkable. If humans—you and I—strive to live and thrive, to leave something of ourselves in our children, then no less do the plants and animals. Just as we honor our own life urge, we should honor the life urge of other beings.

By 2060, the same distance into the future as 1970 was to the present, half of Earth's *species* will have gone extinct. Far more than half the frogs and butterflies, far more than half the diaphanous creatures of the sea—the sea angels and brittle stars—will go extinct, scientists tell us, unless we take urgent action. These projections are based on current rates of extinction; but increasing ocean acidification, habitat destruction, and toxicity will only increase the rates of dying. And not just the species will disappear; with each species, the branching tree of possibilities that its genome holds in safekeeping disappears forever from the planet.

"Oh, for heaven's sake," some people say. "Change happens. Animals come and go. Evolution is a game with winners and losers. If it weren't for the dinosaur extinctions, there wouldn't be human beings. Hoorah for the fifth extinction." All true. But there is a distinction between change and destruction, and that is the difference between death and murder—the introduction of an agent who does the damage for his own gain and satisfaction, in disregard of or contempt for other creatures' will to live.

And no one should assume that the human species can win this game of winners and losers. We are like people who live in the penthouse of a hundred-story building, Daniel Quinn writes in *Moral Ground*. Every day, we head down to remove blocks from the foundation, so we can make our penthouse bigger, fancier. This might work for a few days. But for thousands of days? At some point, the residents of the tower will have created so many channels of emptiness that the entire structure will collapse. "When this happens—if it is *allowed* to happen—we will join the general collapse, and our lofty position at the top of the structure will not save us."

Ever since I read this analogy, I have been haunted by a nightmare vision of my grandchildren's beloved children. With their arms spread, they are falling from a tower, crying out like broken sparrows.

Here is the Dalai Lama:

> Morally speaking, we should be concerned for our whole environment.
>
> This, however, is not just a question of morality or ethics, but also a question of our own survival. . . . If we exploit the environment in extreme ways, we may receive some benefit today, but in the long run, we will suffer, as will our future generations. . . . When the climate changes dramatically, the economy and many other things change. Our physical health will be greatly affected. Again, conservation is not merely a question of morality, but a question of our own survival.

The language of extinction is too passive to express its horror. "Species go extinct," people say. But the fact of the matter is that species don't always go extinct the way bananas go bad, or bombs go astray, or elderly uncles go crazy. Human decisions sometimes drive animals to extinction. Human decisions extinguish entire species. *Extinguish*: to cause to cease burning. All the little sparkling lives. "Shit happens," we say. And sometimes it does. But the fact of the matter is that sometimes, shit doesn't just happen. Sometimes, human beings deliberately create the conditions under which shit is more likely to occur. That is the case with this terrible dying.

I object also to the language of the sixth extinction and will not use it. The current extinction is something morally different from the first five. For all their horror, for all their calamitous power, the early extinctions were natural Earth processes, what the insurance industry calls "acts of God," beyond human control or culpability. This current great wave of dying is the direct result of human decisions, knowing and intentional, or willfully and wantonly reckless. That's a difference of moral significance. It changes a calamity into a cosmic crime, a perversion of human power, a failure of our responsibilities toward the "beauty of life that is now in our hands," to use Mary Evelyn Tucker's words. To call this just the sixth in a long series of extinction cycles is what philosophers call a "category mistake"; it's not the same thing. Extinctions one through five call us to awe. Number six calls us to rage—rage against the dying.

There is one more issue of language to be pressed. If the human impact on Earth has "cut to the very bone" of deep time, effectively ending the Cenozoic era—the time since the dinosaurs, during which the landforms of Earth have moved into their current forms and the hominid mammals have evolved—then Earth is entering a new geological period. It is going to need a name.

Scientists suggest that something like Anthropocene makes sense. But we should use words cautiously. Words are powerful, magical, impossible to control. With a single misguided phrase, words can move a concept from one world into another, altering forever the landscape of our thinking. It's essential that we get this straight now.

So no, not the Anthropocene. That name completely muddles the message. We don't name new eras after the destructive force that ended the era that came before. The International Commission on Stratigraphy didn't name the time of the dinosaurs the Asteroidic, even if an asteroid is suspected of having ended the Cretaceous. The end-Cretaceous asteroid did its damage and sank into the Earth, and the future sailed on. The period after the Permian isn't the Super Volcanoic. The volcanoes did their work and then unraveled into the oceans. So why would they name the next geological period after what destroyed all that came before? Like the asteroids and the volcanoes, human power will crash into the Earth and sink away.

The next epoch, if it has a name at all, should be named after rock bearing the evidence of *what comes next*. That will not be us.

Given human impacts on the Earth, everything will change—true. But human fossils will not dominate the fossil record for the next geological epoch. Nor will the future be built to a human blueprint. Proud, solipsistic creatures that we are, we can convince ourselves that we *are* shaping the Earth—and, for a blink in time, it may be so. We have drawn perfect lines across the landscape, fencerows parceling out property boundaries and delineating poisoned fields of corn and soybeans. But what we are sowing in those squares are the seeds of the destruction of our proud visions. How long will it take the whirlwinds to sweep them away, and along with them the chances of our children? And now, the very notion that humans have become the "deciders," the managers of Earth, makes Earth guffaw in swirls of violence. If we are shaping anything at all, we are shaping climate chaos, and chaos in the ocean and on the land. If there is a voice in that whirlwind, it is not the voice of man. That deep future will not be the Anthropocene.

So what is the right name for this new geological period?

One tradition has it that periods are named after the place where geologists found the boundary between two assemblages of fossils—the lower rocks holding the squashed remains of the extinct species and the rocks above holding the remains of the species that swept in to take their place. Thus the Devonian period is named after rock layers found near the village of Devon, in England. Okay. If we are to name the next period after a place where the boundary between the rubble of the old era and the new is clearly seen, then perhaps we are entering the *Dubaicene,* for that mirage city built of petroleum.

Other geological periods are named after a characteristic of the dividing layer. Then we're probably in the *Unforgiveablecrimescene*. Or the *Plasticene*, as some have suggested.

And some periods are named after the characteristics of the layers laid down during the period's expanse of time. Thus the Cretaceous period is named for *creta,* chalk, after the extensive chalk fields where geologists dug into the depths of that time. If we name the coming geological period after the layers of rubble that pile up at the end of the Cenozoic era—the ruined remains of so many of the living beings we grew up with, buried in human waste—then we are entering the *Obscene era*. The name is from the Latin: *ob-* (heap onto) and *caenum* (filth).

So. What is a person to think of all this dying? Extinctions are silent and invisible—we don't hear the absence of sound. We don't see what is gone from our lives. Moreover, the cases that tear our hearts—the polar bears on melting ice floes; the golden frogs harvested for their elixirs of potency; the tigers, ditto; the rhinoceros, ditto; the black bears, ditto (what *is* this crying need for potency?); the elephants killed for their ivory—these take place far away, driven by baffling forces.

Environmentalist David Foreman answers: "Our asking should . . . begin on the bedrock that we of course have an obligation to wild things of all species, today and tomorrow, to honor their intrinsic value and thus to act only in ways that keep whole the beauty, integrity, and stability of the Earth. Those who want out of that obligation will have to fully show how it is okay to snuff life for short-term, selfish ends. This shifts the burden of proof in a strong and mindful way. Those who would shatter life will have to show in a deep, wide way why this careless, carefree, uncaring behavior is good. That will be hard—so hard—to do."

In the absence of a cogent defense of killing, we have two sorts of duties, negative and positive—what one must not do, and what one must do. We may not be able to preserve a future for all these marvelous creatures, but surely it's a moral minimum to *stop killing them and destroying their habitats.* That is the ecological analogue of the Hippocratic oath: First, do no harm. Herewith, an ecological oath:

AN OATH FOR THE WILD THINGS

1. I will not buy poisons or introduce them in any way into the world. Nor will I buy food that has been sprayed with poisons.

2. I will not destroy unspoiled habitats—not by building a house on unspoiled land, tearing out a meadow for a lawn, or plowing a grassland for crops.

3. I will not step foot in a Kmart, Walmart, house, or any other building that was constructed on the recently bulldozed remains of what previously was a native habitat.

4. I will not refrain from loudly guffawing when someone says it is possible to mitigate the destruction of a natural place by creating a new one.

5. I will not own an outdoor cat and I will not be afraid to badger my neighbors about theirs, knowing that domestic cats are the leading cause of the death of birds and mammals in the United States.

6. I will not be stingy about giving my support, my money, my time, and my vote to those who create natural reserves—marine reserves, wildfowl reserves, wilderness reserves, public land trusts, urban wilderness parks, and other places where animals can thrive.

7. I will not buy products created by destroying tropical or temperate rain forests—not coffee, not tea, not fine wood, not toilet paper, not cedar shingles or two-by-fours or hamburgers. Nor will I buy fish or other wild creatures that are not sustainably harvested.

8. I will not invest in companies that profit from death and deceit. Let's see: ExxonMobil, BP, Halliburton, Monsanto, Coca-Cola. Perdition, the list is long!

9. I will not oppose subsistence hunting or other respectful hunting of abundant species for food, but I will howl about trophy hunting, plinking, or any other destruction of animals for "sport," "fun," "father–son bonding," or any other grotesquerie.

10. I will not plant exotic species in my garden—none of the holly and the ivy, no Scotch broom. The land I have title to, city or country, will be native habitat for wild creatures, even if I have to plant it myself.

11. I will not take more than I need from nature's bounty, understanding that what I take is taken from some other creature that has an equal claim to the conditions for life.

12. I will not pretend that a person can be an upright citizen who kills wild creatures in the course of his business, or causes them to be killed, or profits from the destruction of their habitat.

13. I will not allow my church to destroy the divine creations that it celebrates inside the walls. It's time for a new sanctuary movement that transforms the sweeping lawns of churches into true places of safety, bursting with birdsong, protected forever from bulldozers. It should not be possible to walk into church without getting seeds in your socks.

14. I will not worry about being a sanctimonious pain in the butt. I wouldn't worry one minute about trading a friendship for the ongoing existence of meadowlarks. *Someone* has to take a public stand, and if that makes other people feel guilty, maybe it's about time.

NOW, WHAT *can* be done? Portland writer David Oates thinks that Noah had the right idea when he built the Ark. He knew that whatever survived the Great Flood would repopulate the world—the lions and the elephants, two by two. Now, millennia later, the world is going through a biological bottleneck, a pinch point as brutal as God's fury in Noah's time. What species make it through—that's what the world will be made of. I remember that Noah protested: *I'm old, I'm tired, why me, o Lord?* I ask the same thing; hell, I *feel* as old as Noah. The answer is that it's got to be everybody, each asking, "What ark can I build?" What restored marsh, what replanted prairie, what wildflower garden, what butterfly reserve, what organic farm, what beehive, what woodlot, what artificial reef? And this: What destruction can I stop? What oil terminal, what parking lot, what GMO crop, what poison-spraying truck, what clearcut?

"It comes to me that every good act might be an Ark, no matter how small," David wrote to me, "a widening arc of consequence, incalculable. That there will be flotillas of Arks, uncountable. Tiny handmade ones, and massive ones; science Arks like battleships, and garden Arks like rowboats, all set into the forward river of time, to sail if possible through the narrow part of the hourglass of our times. Some by accident. Some by design. There will be Arks for fungi. Arks for megafauna. Grandstand wilderness Arks with million-dollar funding, and anonymous Arks, memorials of lost passion. Arks like Moses adrift in the bulrushes. Arks like coracles of Irish monks, ferrying their manuscripts to safe places. Arks intentional and Arks ironic. Arks like flotsam in a torrent, carrying who knows what. Seeds and flowers, babies and youngsters. Spores and flagellates, dirt clods and lumps indistinguishable. And then what? To touch, chancewise, on dry land. And start the world anew."

Yes, I say to David. *Well said.* We have to do everything we can to stop the flood, or at least to lessen it, to keep it from washing over the mountaintops. That's the first thing and the most important. And then we have to save what we can, fishing it out of the sloshing waves and bringing it to safety.

Here is an analogy that philosophers often use when they are trying to parse out moral obligations: If I am walking along a trail beside a river and I hear a child calling for help, sputtering and splashing, do I have an obligation to rescue him? Let it be stipulated that I did not push him in—that, in fact, I do not know this child. Let it be stipulated that I know how to swim. Do I go in after him? Yes. Must I? Equally, yes. What is the basis of that obligation? That a child's life is a good and beautiful thing, full of promise, and I am in a position to save it. That the child, as I do, wants to live. That I believe in life. That if someone offered me the chance to explain why I did not have to save the child, I would not know what to say.

The lives of the children of frogs and flowers are also good and beautiful things, full of promise. In many cases, humans are in a position to save them. If someone offered me the chance to explain why people do not have to do what they can to save the plants and animals, I would not know what to say. And is the obligation not more evident, given that in most cases, human recklessness has elbowed the endangered species into the river or human greed has hurled them into the flood?

OVER THE YEARS, college students have often come to my office distraught, unable to think of what they might be able to do to stop the terrible losses caused by an industrial growth economy run amok. So much dying, so much destruction. I tell them about Mount Saint Helens, the volcano that blasted a hole in the Earth in 1980, only a decade before they were born.

Those scientists were so wrong back in 1980, I tell my students. When they first climbed from the helicopters, holding handkerchiefs over their faces to filter ash from the Mount Saint Helens eruption, they did not think they would live long enough to see life restored to the blast zone. Every tree was stripped gray, every ridgeline buried in cinders, every stream clogged with toppled trees and ash. *If anything would grow here again,* they thought, *its spore and seed would have to drift in from the edges of the devastation, long dry miles across a plain of cinders and ash.* The scientists could imagine that—spiders on silk parachutes drifting over the rubble plain, a single samara spinning into the shade of a pumice stone. It was harder to imagine the time required for flourishing to return to the mountain—all the dusty centuries.

But here they are today: On the mountain, only thirty-five years later, these same scientists are on their knees, running their hands over beds of moss below lupine in lavish purple bloom. Tracks of mice and fox wander along a stream, and here, beside a ten-foot silver fir, a coyote's twisted scat grows mushrooms. What the scientists know now, but didn't understand then, is that when the mountain blasted ash and rock across the landscape, the devastation passed over some small places hidden in the lee of rocks and trees. Here, a bed of moss and deer fern under a rotting log. There, under a boulder, a patch of pearly everlasting and the tunnel to a vole's musty nest. Between stones in a buried stream, a slick of algae and clustered dragonfly larvae. *Refugia*, they call them: places of safety where life endures. From the refugia, mice and toads emerged blinking onto the blasted plain. Grasses spread, strawberries sent out runners. From a thousand, ten thousand, maybe countless small places of enduring life, forests and meadows returned to the mountain.

I have seen this happen. I have wandered the edge of Mount Saint Helens's vernal pools with ecologists brought to unscientific tears by the song of meadowlarks in this place.

My students have been taught, as they deserve to be, that the fossil-fueled industrial growth culture has brought the world to the edge of catastrophe. They don't have to "believe in" climate change to accept this claim. They understand the decimation of plant and animal species, the poisons, the growing deserts and spreading famine, the rising oceans and melting ice. If it's true that we can't destroy our habitats without destroying our lives, as Rachel Carson said, and if it's true that we are in the process of laying waste to the planet, then our ways of living will come to an end—some way or another, sooner or later, gradually or catastrophically—and some new way of life will begin. What are we supposed to do? What is there to hope for at the end of this time? Why bother trying to patch up the world while so many others seem intent on wrecking it?

These are terrifying questions for an old professor; thank god for the volcano's lesson. I tell them about the rotted stump that sheltered spider eggs, about a cupped cliff that saved a fern, about all the other refugia that brought life back so quickly to the mountain. If destructive forces are building under our lives, then our work in this time and place, I tell them, is to create refugia of the imagination. Refugia, places where ideas are sheltered and encouraged to grow.

Even now, we can create small pockets of flourishing, and we can make ourselves into overhanging rock ledges to protect life, so that the full measure of possibility can spread and reseed the world. Doesn't matter what it is, I tell my students; if it's generous to life, imagine it into existence. Create a bicycle cooperative, a seed-sharing community, a wildlife sanctuary on the hill below the church. Raise butterflies with children. Sing duets to the dying. Tear out the irrigation system and plant native grass. Imagine water pumps. Imagine a community garden in the Kmart parking lot. Study ancient corn. Teach someone to sew. Learn to cook with the full power of the sun at noon.

We don't have to start from scratch. We can restore pockets of flourishing lifeways that have been damaged over time. Breach a dam. Plant a riverbank. Vote for schools. Introduce the neighbors to one another's children. Celebrate the solstice. Slow a river course with a fallen log. Tell stories of how indigenous people live on the land. Clear the grocery carts out of the stream.

Maybe most effective of all, we can protect refugia that already exist. They are all around us. Protect the marshy ditch behind the mall. Work to ban poisons from the edges of the road. Save the hedges in your neighborhood. Boycott what you don't believe in. Refuse to participate in what is wrong. There is hope in this: an attention that notices and celebrates thriving where it occurs, a conscience that refuses to destroy it.

From these sheltered pockets of moral imagining, and from the protected pockets of flourishing, new ways of living will spread across the land, across the salt plains and beetle-killed forests. Here is how life will start anew. Not from the edges over centuries of invasion; rather, from small pockets of good work, shaped by an understanding that all life is interdependent, and driven by the one gift humans have that belongs to no other: practical imagination—the ability to imagine that things can be different from what they are now.

OUR DESCENDANTS WILL want to know how it feels to live during the time of extinction. I wrote this essay for them.

THE FROGS (MOTET FOR FORTY VOICES)

Deer sleep silently on this land. They choose the quiet edges under apple trees, or the shadowed places where Douglas fir boughs sweep into deep grass. I too sleep in this old Oregon field, now and then. Like the deer, I make my oval bed by pressing down the tall fescue and Queen Anne's lace. Then the space is encircled by a palisade of stems, and I feel as hidden and attentive as a fawn.

One night in October, I tucked my sleeping bag into a hollow beside the swale and settled in. Night was clear at first, warm and shiny black. But when the temperature dropped, the air thickened in an instant. I slept at the bottom of the sea, the air was that wet and heavy. Morning light didn't splash into the eastern sky. It grew out of the moisture, every droplet twinkling, the very air brightening, and that's when the birds began to sing.

The first was the robin's timid question. *Verily truly?* Not really sure, she was wondering, the way small children wonder, lifting their heads from the pillow to whisper, *Is it morning?* The robins were still testing their surmise that a new day follows every night. I lay quietly in the hollow. *Verily truly?* When I woke again, the air was alive with twittering and a sort of slurp that I knew were swallows. And then the meadowlark sang.

The song of the meadowlark seemed as eager and exultant as the morning. Why wouldn't he have been glad? He was singing from a bone-deep confidence that when morning warmth landed at the far edge of the field, the mist would rise from the tarweed. Insects would rise too, midges and mayflies, ascending into the sound of their rustling and the distant-surf roar of traffic beginning to move on Highway 34. It was the promise of a day of bustling light and flies—a good day for a meadowlark.

How did the meadowlark's song sound? Like a flute underwater. Like light on a riffle. Like dampness on a yellow squash. A pure line of song, softened somehow, and slippery. My bird book said the words to his music were *sleep loo lidi lidijuvi,* but not this meadowlark. On one side of me, *Sleep, and wake in the morning.* Silence and a flutter. Then on the other side, *Sleep, and wake in the morning,* as if the meadowlark were lofting reassurance into the air like milkweed seeds from a split pod. How did I feel, to hear that sound? I felt I was alive in a world I could trust. I felt that confident of the morning. A birdsong could do that, could fill a heart with comfort.

I knew that although the meadowlarks left the fields in late autumn when the leaves left the oaks, they always returned in early spring, when small leaves began to bud. I knew that about the time the meadowlarks headed out, geese would begin to call, shotguns would ring against the hills, trains would sing out, great horned owls clack and moan, and in a few weeks the tree frogs would begin to sing. It always happened that way.

At first, a frog would sing alone, a croak from the nearby marshland. Then he was joined by a croak from the far side of the field. Then, around the oval of the pond, voice after voice joined the chorus. Not just forty: a thousand, two thousand, a massive welter of sound. And gradually the frog voices came together somehow into a shared rhythm, a quick, steady pulse.

When I first heard this pulse, I held my breath and listened closely. The frog chorus was as rhythmic as waves on a beach. Were the frogs singing from a shared score? Were they conducted by some unseen hand? Were they resonating with the rhythm of some larger, some cosmic song?

Scientists say that after his call, a frog pauses to suck in a deep breath and then hurries to be first to call again. As the frogs race each other, their voices organize themselves into a resonance, a pulsing wave of sound. When I listen in the night, the whole universe seems to sing in 4/4 time. The cicadas do this too, with their cadenced rasps.

It's possible that creation worked this way as well, breathing life into the rhythmic echoes of repeated refrains. Life began with a simple basic pattern and *repeated* it now, again, *replicated* it now, again, *complicated* it now, *connected* it, again, into the lyric chords of all creation. Catalytic RNA and the musical form of the motet display the same powerful, complexifying beauty.

For a while humans lived as part of the antiphonal progress of the harmonies and the seasons. I imagine people found confidence in Earth's patterns, anticipating the returns and departures, the rotations and revolutions, the melodies, the swelling songs.

Thomas Tallis, an English composer, wrote *Spem in Alium* ("Hope in Another"), a motet for forty voices, for the Elizabethan court in 1570. Did he listen at the open window as the seasonal music unfolded? Did he hear autumn leaves sing a clattering chord when a gust of wind blew through? As he watched the oaks turn scarlet that autumn, Thomas might have heard the reed warbler's twitter and trill. The day he died, November 23, 1585, the nightingales would have been gathering their own strength for migration, and the terns, the redstarts and swallows. It must have been a crowded day in the singing sky, the day his spirit rose. Did you see this, old Thomas—pirouetting flocks of tiny birds, weaving and folding, specks of light swirling into cyclones and bursting into chrysanthemums, birds spreading into chords, appearing and disappearing as if they had no substance at all?

Maybe his faith in God was lifted by the assurance that birds would go and birds would return. That storms would come in season and storms would blow back to sea again, that lilies would bloom, grow hard with seed, and bloom once more. The music of the world was a repeating promise, a promise that harmony would be restored again and again in chords so complicated and beautiful that we could love them, even if we could not fully understand the genius of their pattern.

Even until just a few years ago, birds came and went so reliably that scientists made charts as precise as scores to predict their travels. In Oregon, the first rufous hummingbirds returned in late February when the blueberries bloomed at the coast. Violet-green swallows returned to their ponds in early March, when mayflies emerged. It was a great day in the swamps in early April, when both the American bitterns and yellow-headed blackbirds swooped in, grumping and hollering. The humans and the birds slept and woke by this, lived and died by this faith in inevitable, unfolding harmony, the expectation and the arrival, the call and the response, the question and the answer, the world's promise of absolution and return.

FOR SOME TIME that morning, I followed the meadowlark through the field. At first, he sang from the fencepost beside the oaks. But by the time the sun finally flared over the distant volcanic peaks and spiked into the steaming field, the meadowlark must have been hungry. He set about flittering, sailing, flittering, sailing, dropping into the grasses to capture, I suppose, grasshoppers and to grub around in the dirt. I did not hear him sing again that day. What I heard were the alarm calls of Steller's jays, the flocking calls of crows, and the whispered whistle of one varied thrush. Earth sailed in its grand orbit around the sun, which swept a long arm over the planet, cueing first the sopranos, then the altos and the thrushes.

The pleasure was in the successful prediction: Yes, it won't be long until huge flocks of Vaux's swifts will vortex down the chimney of the food science building on campus. With our children, we lay on our backs in the parking lot at dusk to watch the spinning flock. The flock leaned over the chimney and spilled a hundred birds into its maw, and then righted itself and spun away, only to return, and return, until all the birds had been poured in for the night and it was bedtime for the children too.

Because the children had well-founded faith in the regularity of the comings and goings, they loved surprises—an unexpectedly warm day in February that caught even the azaleas off guard, a thunderstorm in a Pacific valley that rarely got thunder, an early cinnamon teal in the snow. They were the exceptions that proved the rules, the aberrations against which we measured the norms, a childlike glee to see even the world breaking the rules, which meant that there *were* rules, there was a way the world worked that we could see and to some extent understand. And so, once every two years or so, we stood at the windows as lightning bedazzled the night and scorched the transformers, and we called our friends to marvel. In the morning we reset the clocks, sure that they would keep the same time as before the storm.

People in my town kept track of this—the first vulture, the first bluebird, the first rose, the first blueberry. It was always the first arrivals they wrote on their calendars. It's impossible to keep track of the last, because you never know it's the last until it's gone forever. The last meadowlark song, the last bloom of the Mr. Lincoln rose—who knew if tomorrow would bring another, or the next day another, until finally they did not come again? But, dear Thomas, things are changing.

The weather comes now and goes, and who can make sense of it? Cold places are warm, and warm places are cold. Rainstorms and droughts are relentless. Last year, drought dried the grass fields where insects and mice would have grown abundant, and the peregrine falcons, skinny or starved, produced no young. This year the swallows came back to Oregon before winter was finished. They followed the coast, where the weather is usually mild and mayflies are hatching in the ponds. But a terrible late-winter storm blew in and there were no insects in that wind. Have you seen a starved swallow, swinging by its little claws, upside down from a wire? Have you seen the frozen eye of a swallow? It's white, Thomas. You can't see into it.

But it's not just the violence of the weather, it's the treachery in its cues. Migrating birds mostly time their departure by the length of the day. But plants and insects generally respond to the temperatures. These used to coincide. In the great sweep of Earth's baton, the buds and insects emerged just as the birds entered. But some of them don't arrive at the same time anymore, not exactly, now that the world is warming, and this is where things break apart. You can't make a small change in the time that the tenors enter and expect the music to work at all. You can't make a small change in the pitch of the sopranos and expect harmony. You can't make a small change in the temperature that cues insects to emerge and still expect meadowlarks to sing. It's all one thing, this motet.

And what will the hummingbirds do, when they arrive after the salmonberries have finished blooming and the blueberry blossoms are spent? And the great tits in your country, Thomas? The birds enter the forest with exquisite timing, laying their eggs so they will hatch at the precise peak of the caterpillar population. There are twenty-four days when a newly unfurled oak leaf is palatable to the caterpillars, twenty-four days when the great tits can find enough food for their insatiable nestlings. Less food for them now, when they arrive too late. The choral conductor can tap her baton against the music stand to stop the choir. "Let's try this again, tenors. You see that you are to enter a measure after the altos. I'll cue you." And they can rehearse this. They can sing it again until they get it right. But there is no conductor of the morning choir, and there are no rehearsals. There is just that perilous performance on the very day that the tiny nestlings hatch, when either the weather warms enough to cue the caterpillars or it does not.

Music is a precarious relationship of sounds, moving through time—ephemeral, forever vanishing. That's what a healthy ecosystem is. That's what a life is, a temporal harmony. It may be true that everything beautiful flies apart some time or another; everything flies apart. But music sometimes holds it together; that is the miracle in music. Meadows sometimes hold it together; that is the miracle of meadows, with all the parts in their skittering, soaring places. And aren't people trying to hold it together, each in our own way? Dear god, aren't we trying?

Even in our grief, we know that change works like frog songs and cicadas. It organizes itself. Like an ocean wave that rides up the back of another wave, like waves of sound that overlap and suddenly double in amplitude, like the swelling chorus of the choirs singing together, change can suddenly become something much larger and more sudden than anyone expects.

This past autumn, I visited the Cloisters museum on the wooded cliff edge of the Hudson River in Upper Manhattan. There, within a damp palisade of stone, forty speakers on tripods were arranged in an oval. Each speaker was ready to send a single recorded voice into the empty air, to re-create Thomas Tallis's motet for forty voices. I stood in the center of the space. At first, there was only the lap of shoes on stone, like leaves slapping in rising wind. Then a voice as clear as a thrush sang out, and another. People held still and began to listen. They closed their eyes and listened. Some dropped to their knees and listened. Listening made the oval space a suddenly holy place. The music piled up, more voices, richer harmonies. It filled our bodies. *Domine Deus, creator caeli et terrae.* The silence of the listeners, the forty voices joining, one after another, into perfect chords that lifted and banked like a flock of swallows—I cried. Many people cried.

I believe that everything awaits redemption at every moment. Redemption: when the pieces, which are scattered, are brought together again into the vibrant harmony of living systems, when the whole that they create is beautiful and embedded in even greater miracles of relation. We yearn to be called back in. Everything yearns to be called back into a right relationship, the frogs into their chorus, the cicadas into their pulsing choir, the people into Earth's harmonies, the dancing insects into their light.

the rights of nature

WHAT DOES EARTH ask of us? Restraint and respect, at the very least. But maybe that's not enough. Maybe it's time to write restraint and respect into our laws, by recognizing the rights of nature. What would that mean? What moral sense does it make? How is it playing out in Ecuador, the first nation to grant constitutional rights to the natural world? I went there to find out. Here's my report:

PACHAMAMA'S ANGER

On Española Island in Ecuador's Galápagos archipelago, all the animals are crying for their mothers. Sea lion pups bawl. Nazca boobies honk and whistle. Swallow-tailed gull chicks do some aggressive, clattery thing. Only the lava lizards are silent as they leap from rock to rock. If they are hissing, I can't hear them over the belch and squawk. And the smells? Guano, above all else, and the malodor of a desiccating storm petrel, splayed on the gravel. From the turds of sea lions, the smell of dead fish and sick dog. A booby stands on the downy corpse of an oversized chick, kneading the body with its greenish feet. Whether this is clumsy CPR or blind rage or some gruesome king-of-the-mountain game, I do not know. The lava itself exudes the sulfur and soot of creation. Some plant smells like a sweet pickle. I smell heavily of coconut sunscreen and sweat.

Every little island in the chain has its little cloud, and every little cloud has its island, and that's apparently the rule: If a cloud has no island, it is obliged to vanish. The lava mounds are black and red, and so are the marine iguanas that hang vertically from claws in rock crevices, their tails in the sea. Every rock has some gaudy animal standing on it and something white dripping over it, and there are many, many rocks. A sea lion arches his back and raises his head, swaying. There is no mistaking his place in the hierarchy of living things.

As for me, I know my place too. I mean this figuratively and literally: I have been put in my place here among the wheeling lives. It's quite a small place beside a white post, which is the only place I am allowed to sit. If I venture too close to a nest, a quick peck from a booby will remove a divot from my calf, and a park ranger will gently admonish me, "Stay on trail, please." I'm not in charge here. I have no right to rummage around or to take so much as a pebble. The animals will allow me to sit quietly for a short time, but the rights holders in this place are the marine iguanas and the baby sea lions and maybe a couple hundred milling shearwaters. In this place, I am required to show respect: Talking baby talk to the animals or demeaning them in any other way is expressly forbidden.

I don't have much moral standing in the park either. I have instrumental value to the island, I suppose. I am tolerated for the modest ecosystem services I provide: a slug of shade now occupied by an iguana, and maybe a bit of moisture if I were to pee (also forbidden). My corpse wouldn't have much value; there are no big scavengers here, and there is plenty of calcium already. Certainly I have economic value—U.S. *dólares* for the guides and guards who keep nonnative species like me at bay. But I can't imagine that I am regarded by the animals as having any intrinsic value at all. Certainly not for my beauty: I am squatting here in baggy shorts, wearing a striped bath towel folded on my head. I shift my towel to hide from the ferocious equatorial sun. Another iguana sashays her great weight into my shadow. A swallow-tailed gull shoots out an arc of guano. It is very, very hot, even this early in the day.

Under ordinary circumstances, I would resent being relegated without a vote to subsidiary status. I'm an adult *Homo sapiens,* for god's sake, the pinnacle of creation, and I'm not used to animals using me and showing me this complete disregard. Before the day is out, a booby will walk right over my feet and never even care. In the Galápagos National Park of Ecuador, the animals are free to do as they please. I have to follow strict rules for their benefit or be banished, while their rights to enjoy life and liberty and to pursue happiness in their own lunatic ways are absolute. I've always wondered what it would be like to live in a world that was flipped on its head, if our species had no rights, while other species had the right to walk all over us without regard or consequence. Now I know. It makes me strangely happy.

North American conservationist Aldo Leopold wrote that the history of the moral evolution of humanity is mapped by the history of the expansion of the sphere of our moral concern. As humanity has become more ethically developed, he said, people have embraced a wider and wider circle of beings as worthy of moral consideration and thus worthy of rights—first family members, then clan members, then property owners, then former slaves, then women, and so it goes. If he is correct, then human beings have just made a quantum leap forward in moral development. That's because Ecuador in 2010 became the first nation on the planet to grant constitutional rights to the natural world—to Pachamama, who is Mother Earth, nature, the natural world of plants and animals, volcanoes and waterfalls, lizards and hummingbirds. Sixty-five percent of voters approved the change.

The enumerated rights of the natural world are extensive in Ecuador. They effectively grant natural objects legal standing as persons. The rights holders are species, ecosystems, and natural cycles, including cycles of regeneration and evolution. The rights are held directly: It is damage to the natural systems and to their members—to their existence and regeneration—that matters, not the damage to the human beings who benefit from their natural processes. The rights are both negative and positive. Negative rights *prohibit* actions that disrespect and damage the natural world; positive rights *require* government action to prevent harm. Here is some of the language in the Ecuadorian constitution:

> ARTICLE 71. Nature, or Pacha Mama, where life is reproduced and occurs, has the right to integral respect for its existence and for the maintenance and regeneration of its life cycles, structure, functions and evolutionary processes. All persons, communities, peoples, and nations can call upon public authorities to enforce the rights of nature . . .
>
> ARTICLE 72. Nature has the right to be restored. This restoration shall be apart from the obligation of the State and natural persons or legal entities to compensate individuals and communities that depend on affected natural systems . . .

> ARTICLE 73. The State shall apply preventive and restrictive measures on activities that might lead to the extinction of species, the destruction of ecosystems, and the permanent alteration of natural cycles . . .

I would say that this is a very big change. Others say that these are just words—and real change will not come until the words are put into action. A young, open-faced Ecuadorian friend, Orlando, is one of these people. "Yes, yes," he says, "but this is just for the indigenous people. No one pays attention." But to a philosopher (as I confess to being), this is a sea change, a paradigm shift, a new beginning, a testament to the human ability to see the world in new ways, even if (maybe *especially* if) these are only words. Words express ideas, and these particular ideas are powerful and consequential. They tell a new story about who we are in relation to the world, we humans, and how we ought to live.

When I asked another Ecuadorian friend, Francisco, about the importance of the rights of nature, he led me down a trail that dropped steeply into a deep and shaded river gorge (entrance fee, *un dólar*). It worried me that at six thousand feet in the Andes, every step down was going to have to be a step back up, but there were moss-roofed ice cream stands now and then, each with its Donald Duck litter barrel, so on down I went. Water dripped from lianas and pooled in bromeliads. The trail was moss-green cobbles, slick enough to require a rented walking stick (*cincuenta centavos*). Beneath fern-covered slabs of concrete dislodged by an earthquake ("Pachamama can be angry"), under a carved sun and moon ("the children of Pachamama"), through a sweet cloud of orchids ("Pachamama is very beautiful"), past a Coca-Cola booth (*un dólar*), we followed the path toward the sound of thunder ("*Por favor,* a plastic bag for your camera?").

Around the corner was a stupendous waterfall. It rumbled over a ledge, bashed into a giant basalt tooth, and pulverized into mist that filled the cauldron. Two rivers swirled past the tooth. The air was suddenly cold. In fact, the air was suddenly not air but water. Water soaked my raincoat, drenched my face, and wetted my camera, because of course I had turned down the plastic bag. Who can take pictures through a plastic bag, and who could have imagined a river so completely transformed into spray?

There had been plans to build a huge hydroelectric plant at this waterfall. It didn't happen. The local people make their living from this waterfall, Francisco explained—the renters of walking sticks and purveyors of ice cream. They told the government they did not want the waterfall destroyed. Besides, they said, the waterfall embodied Pachamama in her greatest power and beauty, and threatened to embody Pachamama in her greatest fury. The hydroelectric plant went in downstream. Who can say which argument was most persuasive?

It's significant, though, that Ecuador's new constitutional provisions have been upheld in court. The case: *Wheeler c. Director de la Procuraduria General del Estado de Loja*. The issue: Can a provincial government's road-widening project dump construction debris into a river, narrowing it and creating floods that erode downstream lands? People brought the suit, of course. But here's what's new and huge: They filed suit for violation of the rights of the land itself, complaining of damage not to themselves but to the riverbank for its own sake. Never before has a riverine system been able to sue on its own behalf without being thrown out of court for lack of legal standing. And not only was the voice of nature heard in court—nature won the case.

The court ruled that the right of the river to be whole trumps the right of the local government to damage it. And it ruled that the burden of proof was not on the river to prove that it had been damaged; rather, the burden was on the government to prove that its earthmovers had done no harm. Once again, it's possible to say that nothing has changed, because the local government has not yet removed the rocks or repaired the river. I would say that, on the contrary, this precedent offers an important model to a world that has struggled (and perhaps stalled) in its effort to think in new ways about the relation of *Homo sapiens* to other living things and natural systems.

WAY BACK IN 1972, the U.S. Forest Service granted Walt Disney Enterprises license to develop an eighty-acre ski resort in the Mineral King Valley, "an area of great natural beauty" adjacent to Sequoia National Park. In *Sierra Club v. Morton,* the Sierra Club sued to stop the development, arguing that its members, who used the forest for recreation, would be harmed if the forest was harmed. The court threw the case out, arguing that the Sierra Club could not prove significant damage to its members. Nobody talked about the damage to the forest, which would be cut to stumps. Nobody talked about the damage to the rivers or to the salmon, suffocating with silt in their gills, or about the damage to the bulldozed mushrooms or nesting wrens. Nobody talked about the silencing of the frogs. The only harm that counted was to Sierra Club members in their safari vests and binoculars. And that wasn't harm enough.

What the heck was that about? The background is that in the United States, only *persons* have legal standing to sue on their own behalf. That's what it means to be a person under the law—to be able to bring a case in your own name, claim harm to yourself, and use the awarded damages to repair the harm to you.

U.S. courts have no trouble imagining that a corporation is a person and according it rights—even the right to free speech, which is to say, the right to influence elections. But when it comes to animals or plants or forests or natural cycles, or Mother Earth, the courts are unable to embrace them within the circle of moral or legal concern. And so nature has no legal standing, no personhood, no rights.

The *Morton* case stuck in the craw of U.S. Supreme Court justice William O. Douglas—a man of the great Northwest forests. In his famous dissent, he made a case for the legal standing of natural objects. What if the court "fashioned a federal rule that allowed environmental issues to be litigated . . . in the name of the inanimate object about to be despoiled, defaced, or invaded"?

> Inanimate objects are sometimes parties in litigation. . . . So it should be as respects valleys, alpine meadows, rivers, lakes, estuaries, beaches, ridges, groves of trees, swampland, or even air that feels the destructive pressures of modern technology and modern life. The river, for example, is the living symbol of all the life it sustains or nourishes—fish, aquatic insects, water ouzels, otter, fisher, deer, elk, bear, and all other animals, including man, who are dependent on it. . . . The river as plaintiff speaks for the ecological unit of life that is part of it.

Legal philosopher Christopher Stone followed up with his brawny little book *Should Trees Have Standing? Toward Legal*

Rights for Natural Objects. If courts can appoint guardians to protect the rights of infants, giving voice to the grievances of those who can't speak for themselves and protecting them from harm, Stone wrote, why can't they do the same for forests, which also are voiceless but also sentient, beautiful, and full of possibility? But nothing really came of his argument. The United States remained a country where environmental regulations are generally not drawn to protect the natural world from harm. On the contrary, environmental regulations and the Environmental Protection Agency aim to regulate the times, places, and circumstances wherein environmental destruction can take place—to the continuing diminishment of the natural world.

"Ah, but the Endangered Species Act," people protest. In effect, the Endangered Species Act (ESA) grants plants and animals a sort of right to life—once they are teetering on the edge of extinction. It's true that under the ESA, the lives of plants and animals can sometimes trump the economic interests of those who would destroy them, their habitats, or their evolutionary processes. The ESA may not be cast in the language of rights and legal standing, but it certainly can delay economic projects and sometimes stop them in their tracks. An example is the endangered marbled murrelet, whose dependence on ancient forests has slowed the transformation of five-hundred-year-old trees into pulp.

But I'm not the least convinced that the ESA honors the rights of plants or animals. Does it respect their rights to life when the government waits to protect a species until the last possible moment before extinction? Or when it saves the smallest possible portion of habitat? The Endangered Species Act is the stingiest, the most miserly and grudging, the most last-ditch of all possible ways to respect the natural world. No action can be taken until some agency pronounces the situation calamitous. In what way does this respect the rights of any but the tattered remnants of the species? And even if the ESA manages to protect "your tired, your poor, the wretched refuse of your teeming shore," it doesn't even try to protect "the huddled masses." What of the great herds of buffalo, the loping wolf packs, the swirling flocks of trumpeter swans, the migrating hawks, monarch butterflies by the millions, schools of silver salmon—the great *abundance* of lives, the wonder of their numbers? We eat away, eat away, eat away at them until they are almost gone, and then we congratulate ourselves as enlightened for saving the stragglers.

This isn't the way rights are respected. Would it respect human rights to declare "that all men are endowed by their Creator with certain inalienable rights, and that among these are life, liberty, and the pursuit of happiness, *as long as* these men are the last three, or thirty, or three thousand of their species left on Earth. Until then, men are not endowed with anything"? The point about a rights claim is that it gives effect to certain value claims about a class of beings: that they have value not as property or as means to others' ends but in themselves and for their own sakes; that their thriving is thus of overriding value; and that consequently, interfering with their thriving is a violation of moral and legal duties of restraint.

North American philosophers ask, When do species merit rights? Only when the species closely resembles human beings in significant ways, they answer—the ability to feel pain, for example, or the ability to reason or speak or regret, or possession of a soul. The question they don't answer is, Why should humans be the measure of merit? One might think that two-legged creatures who possess pain receptors and reasoning brains and talking tongues and maybe even souls might be able to honor creatures whose abilities are perhaps different but in many ways superior to their own. If they noticed. One might think that humans would realize that the truly morally significant ways in which all creatures are the same is that we all strive to live, and that all lives unfold in complex interdependence.

I'M NOT A particularly brave person, but I want to snorkel alongside a lava cliff in the Galápagos Islands, so I heave myself out of the boat. Right there, right next to me in the water, flitting like flattened ballerinas, are four spotted eagle rays. I plug in my snorkel and swim along behind them. Below me, a school of glittering cardinal fish expands and contracts, as if it were breathing. A couple of dozen Moorish idols swim past the cliff face, all finning in one urgent direction. Sergeant majors. Groupers. Giant damsels. The sea is full of life. I know that's a cliché, but this sea is Full. Of Life. The sea is *made* of life: clouds of krill and a couple of pirouetting sea lion calves, some small orange things, a vertical arrow of silver bubbles that unfolds into a fish-hunting booby, blue-green-pink parrotfish as brightly scaled as parrots themselves. A penguin splashes through bubbles and swims right past me: a tiny penguin with tiny wings. My god. Who would have thought? There are whitetip sharks sitting on the bottom, heavy, as if they had swallowed lead shot. I see the paddling feet of a pelican. This is astonishing—a world completely hidden from any person afraid to jump.

This is a different point of view, from inside this sea of living things, from this swirl of creative energy, generative systems of generative systems. There are hierarchies of size and power in the splash and struggle, but there is no hierarchy of value. Each thing is worthy. Each fractal layer is necessary. If there is a striving in the confusion, it is to continue to live. Life itself is worthy, and so all the small lives are worthy. Floating on my stomach, looking deep through clouds of living things—never have I seen as complete a repudiation of the idea that human beings are separate from the rest of creation, that we are better somehow, that we are in charge, that we are the point of the whole thing.

The moral center of the ancient cosmology that recognizes the rights of nature is the ultimate and absolute worth of nature—the source of all life, the creative genesis, the mother, Pachamama, Mother Earth. Pachamama sings with the wind. She erupts with the volcano. She trembles in the earthquake. She grows with the corn. She is the adobe house. She is the woven cloth. She is the basket of eggs in the marketplace and the basket of chickens. She is the children. She is the spawn of fish. There is no separation between nature and culture, animals and humans—we are all one interactive system, and it is a beautiful, generative, and fearsome complexity with creative, and disruptive, feedback loops. A well-ordered society, the good life, the *buen vivir*, respects the rights of nature, balancing human and nonhuman interests, and honors practices that celebrate and strengthen the thriving of the biocultural world.

It's hard to predict the outcome of the legal protection of nature's rights, in Ecuador or in the United States, where some municipalities also recognize the rights of nature. But there's a paradigm shift going on, and it would be a mistake to think it's a small thing. A revised set of ideas flips the presumptions that govern the relation between humans and nature, like so:

THEN: Humans have the right to destroy nature, unless they are specifically prohibited.

NOW: Humans have no right to destroy nature, unless they are specifically permitted.

THEN: Only damage to humans can stop a destructive practice. It's all that counts.

NOW: Damage to natural systems can stop a destructive practice. Nature counts too.

THEN: Human well-being is accomplished by separation from and conquest of nature.

NOW: Human well-being is inseparable from the well-being of nature.

I'M SITTING CROSS-LEGGED on the bow of our little boat as it motors across a shiny black sea toward a new moorage in the Galápagos Islands. The sun set some time ago. The moon has yet to rise. The stars are brilliant, but all wrong to my eyes. In the place where the Big Dipper should be, there is a small square of stars—the Southern Cross. Orion is there, but he has wandered to the top of the sky. On the sea under each star is a yellow puddle of light. A swallow-tailed gull hangs in the dark wind directly over our bow, swinging its head from side to side, looking for a flash that will map the path of a startled fish through luminescent plankton.

"Ghost gull," a crewman says, and I am fully prepared to believe him, in this place so unlike anything I have ever seen before. I am prepared too, more fully than I have ever been, to imagine myself an equal member of a community of living things, called to honor the rights of fellow members, called to act with respect for all life, and with a restraint that acknowledges the value of all beings.

I'm thinking about what Francisco told me earlier, as spray from the waterfall drenched us all. There is a sort of cosmic justice at work in the world, he said. Pachamama gives back as she gets. If humans live respectfully, in harmony and balance, earthquakes will still shake the mountainsides and volcanoes will continue to pour lava down the valleys. But Earth will be merciful, and the villages will not be destroyed. If humans live stupidly, cravenly, with reckless disregard for the rights of nature, nature's fury cannot be contained.

In the history of the expansion of legal rights, I would point out to Aldo Leopold, I can't think of an instance in which the people in power just up and granted rights to those who did not yet possess the rights. In every case, the powerless demanded their rights and the powerful granted them grudgingly. The legal rights came first, and the moral development took place more gradually, as those in power got used to the idea that the universe of moral consideration had just grown wider. Sometimes the moral development has taken a very long time.

In a time of climate chaos, humans are beginning to feel the sting, the consequences of ignoring the rights of the Earth. Pachamama's storms and droughts and inundations demonstrate the practical wisdom and the moral wisdom of granting her the rights She merits. That means that humans are called to acknowledge new moral and legal responsibilities toward the Earth—to honor the inherent worth of life, living things, and the natural systems that support them; to restrain their own behavior in order to bring their narrowly self-regarding urges into balance with the creative urgency of the natural world; to accept full membership in the family of living things, an interdependent whole that is beautiful and astonishing and alive.

PART III

a call to witness

breaking the silence

TODAY, THE *New York Times* broke the news that the greenhouse gases already in the atmosphere will force planetary heating past the 3.6 degrees Fahrenheit (2 degrees Celsius) mark: "While a breach of the 3.6 degree threshold appears inevitable, scientists say that the United Nations negotiators should not give up on their efforts to cut emissions. At stake now, they say, is the difference between a merely unpleasant world and an uninhabitable one."

One would think that people would gather on the street corners, glancing at the sky, talking about this news in lowered voices so they wouldn't frighten the children. One would think that they would be on the phone to their families. "Have you heard?" But they aren't. I wouldn't call my daughter, because she's the mother of young sons and when I talk about climate change and the future of the planet, tears fill her eyes in fear for her babies. I would be cruel to bring it up. I don't knock on my neighbor's door, because she would politely change the subject. I would be forcing an unpleasant conversation. I don't call my cousins in Ohio, because they would call me an alarmist and want to talk about football. It's a lonely thing, this climate change.

It's also a stunning thing, that we face climatic changes that will undermine the lives of our children—and very few people are talking about it. I don't know why. Maybe it's the natural aversion to unpleasant subjects. Michael Nelson, in his public talks, says that climate change is the conversational equivalent of poop, and this might be right. Maybe it's because people feel guilty, and nobody wants that. Maybe it's because people are unsure if their neighbors will agree with them, and we are socialized to avoid controversial subjects—politics and religion, and now climate change. Maybe it's because they feel there is nothing to be done, so talking about it would be pointless and worrisome, or even boring. Maybe they really do care more about football.

But most likely it's a variety of what American intellectual Lewis Mumford called a "magnificent bribe." The bargain is that "each member of the community may claim every material advantage, every intellectual and emotional stimulus he may desire, in quantities hardly available hitherto even for a restricted minority: food, housing, swift transportation, instantaneous communication, medical care, entertainment, education"—on the condition, I would say, that they never ask where it came from, or at what cost in human suffering, at what cost to the future, or to what long-term effect. That's the deal: If they ask, they have to turn away from their glittering lives.

Silence can be a terrible thing. It's not just the absence of sound. Silence is a thing in itself, a heavy blanket that lies over the body of conscience, pressing it down, suffocating it. Like a blanket, silence hides what is underneath in the soft darkness. Like blankets infected with smallpox, silence spreads; where there is silence, there will soon be more silence. Silence is therefore the servant of wrongdoing; when you want to capture a hawk, no matter how ferocious, all you have to do is throw a blanket over its head.

But let the metaphor go. We have real experience of the effects of silence: A few years ago, the United States was shocked by two disasters that might have had more in common than was immediately evident—the *Deepwater Horizon* oil gusher under the Gulf of Mexico, and the serial child abuse by Jerry Sandusky, a Penn State assistant football coach. Both were enabled by silence—what I call the Sandusky Syndrome.

THE SANDUSKY SYNDROME

A priest, a Boy Scout leader, or a football coach does terrible things to children, and for years, no one talks about it. Years. No one intervenes. *How does this happen?* we ask ourselves. *How does it happen over and over and over again?*

An oil gusher poisons the Gulf of Mexico, natural gas fracking pumps a secret mix of poisons under farmland, and an entire industry makes vast profits by digging and selling fossil fuels that are disrupting the climatic conditions under which life evolved, threatening terrible harm to children and all future children. And for years, no one intervenes—least of all the fossil fuel industry's indentured politicians. Three televised 2012 presidential debates about the future of the country, several Republican debates in 2015, and no one said one word about climate change. How could that have been?

The only surprise here is that anyone is surprised. The fact is, there's a pattern to these cases that should be as familiar as family by now. Jerry Sandusky benefited from a ubiquitous syndrome in American society, a step-by-step sequence of silencing and silences that protects profitable institutions by sheltering wrongdoers. We can see the same syndrome at work in the business plan of Big Oil. Here are the steps in the Sandusky Syndrome and their echoes throughout the fossil fuel industry:

STEP ONE. The child abuser ensures the silence of his victim either by threatening harm to the victim if he tells or by making the victim blame himself for the harm.

First, the threats: "The Keystone XL pipeline is in the national interest . . . to decide anything less . . . will have huge political consequences," said American Petroleum Institute president Jack Gerard. Then, the self-blame and self-loathing: "We have met the enemy and he is us," people tell each other, even though it's BP—not consumers—that cut corners in the Gulf and created an epic oil gusher.

STEP TWO. The secrecy of child abuse is further protected by shaming, ridiculing, intimidating, or dismissing those who raise early alarms. In Jerry Sandusky's case, janitors who witnessed abuse told no one, fearing they would lose their jobs.

Ridicule? Al Gore is "delving into the land of tin foil–lined baseball hats and Martians speaking to us through television static." Intimidation? Boat captains working the BP cleanup effort after the *Deepwater Horizon* oil gusher in the Gulf of Mexico told reporters they had seen large areas of surface oil off the delta, but they would not give their names for fear of losing their jobs.

STEP THREE. When officials learn of a case of child abuse, the first response is to clamp on a lid of secrecy. ("We cannot comment because of ongoing investigations," said officials at Penn State.)

After the Gulf oil disaster, one of BP's first steps was to co-opt or muzzle scientists. In some cases, BP tried to put entire university departments on payroll. Scientists' contracts with BP say they cannot publish the research they conduct for BP or speak about the data for at least three years.

STEP FOUR. When the terrible facts of a pedophile's actions finally come to light, people are astonished: He is an ordinary, giving man who went out of his way to be kind to children.

And "you can be sure of Shell." And Gulf Oil is "a name you know, a name you can trust." And Arch Coal plays "an integral role in supplying a safe, responsible, life-enabling and world-developing resource." Such benevolent corporations: How could anyone have foreseen the devastating effects of their business plans?

STEP FIVE. Institutions deny that there is a systemic problem, admitting to just one rogue actor, one negligent administrator. They take responsibility for not preventing that particular abuse but take no responsibility for creating a system of silence.

The Penn State board of trustees accepted "full responsibility for the failures that occurred." Board members vowed never to let similar abuse happen again. The board might have taken the words straight from BP president Tony Hayward: "To those affected and your families, I am deeply sorry. . . . We know it is our responsibility to . . . do everything we can so this never happens again."

STEP SIX. The victim, exposed and betrayed, struggles against long odds to put his life back together. He will never be the same again. Agencies establish hotlines to report child abuse: 1-800-END-HARM.

Hotline to report Earth abuse: 202-456-1111. The phone will ring in the White House.

❖

THE WRONGS AGAINST the planet call us to witness, to speak the truth, to put into words the horror and the sorrow, to talk about a better way. This is not easy. The role of the witness in history makes clear the costs of telling the truth to people who are not ready to hear it. Consider the story of Cassandra.

Cassandra, a lovely young girl, agreed to marry Apollo, but only in exchange for a great gift: knowledge of the truths of the past and the future, and how to raise the dead and how to heal the sick. Besotted, Apollo promised. But then Cassandra changed her mind about the marriage. "Sorry." Enraged, what was Apollo to do? He could not break his promise, but he could add a codicil. Cassandra would know the truths of the past and the future, but no one would believe her. So off Cassandra went to the Temple of Delphi, where she slept beside a hole in the center of the marble floor. Snakes slid down the hole and crawled into the underworld, where they learned the great truths. In the night, they slithered back into the temple and whispered to each other. Lying awake in the temple, Cassandra listened, and so she learned the truths from the center of the Earth.

The next time we see Cassandra, she is standing at the gates of Troy, howling and tearing her hair, insane with grief, while the townspeople jeer and throw stones. She knows what awaits as the Trojan Horse approaches, but no one pays any attention. In the end, she is taken as a concubine by the great Greek commander Agamemnon and murdered by his wife—an end she surely foresaw but could not prevent.

The logic of Cassandra's quandary is telling. "Don't be such a Cassandra," people say, meaning, "Don't keep howling about doom." But what if the invading army *is* at the door? What if the arctic ice sheets *are* melting? What if the oceans *are* souring and heating, and coral reefs *are* dying? Nobody wants to be the one with the bad news, least of all Cassandra. But Cassandra had no choice; truth-telling was her curse. And do each of us have a duty to warn? The law says so. As a homeowner, I have a duty to warn even trespassers of the hazardous well in my backyard. As a products manufacturer, I have a duty to put a sign on a bucket, warning that small children might tip in and drown. As a therapist, I have a duty to violate the confidentiality of my patient if I have reason to think he is dangerous. As a citizen, I have a duty to yell, "Watch out" if I see a child dart in front of a car.

But it's not about the law. Who could stop themselves from yelling a warning when a child is in danger? It's an embodied response, a response not of duty but of humanity. It presupposes not inevitable doom but hope for a better outcome—confidence that it doesn't have to end this way.

THE DUTIES OF THE MOON

We had been talking over tomato soup, three friends. From the kitchen, we looked through open doors onto a small orchard—on one side, the creek; on the other, a hillside of Douglas firs. Late fall. Red apples on the trees, yellow maple leaves on the matted grass. The low sun cast a jagged shadow over the meadow, a silhouette of the clearcut ridgeline to the west. We were serving pie, a good pie, cranberry-pecan pie at that time of year.

We are not young women. We carry our years on thick yellow bones. We have grown our hair long. Our feet are wide. Our knees crackle like flames. The tides no longer ebb and flow in our bodies. For many years, we have looked straight at the Earth and have not turned away. But we felt fear, a bodily alertness to the real danger of the dark days of Earth's time—that they could overwhelm us, pin us down, like the shadow of a terrible bird.

"Do you think," asked Alison, "that the Earth is afraid too?"

Earth, afraid. The sea-surf cringing. The great cedars cowering. The white gulls flattening themselves against the sand.

We could not help ourselves. We began to cry. We pressed our hands over our mouths and wept silently, so the Earth would not know that we were afraid too.

After a time, an apple fell through its branches to the ground.

The creek hushed itself.

There was pie.

Robin filled the teakettle.

The faded afternoon moon cupped a pear tree in its crescent.

We walked, then, out the narrow track along the creek. Maple leaves carpeted the mud, and spent nettles bent over the trail. We passed under red cedars as tall as the sky and alongside a grove of newly planted cedars. There are good and convincing reasons to despair. What we have to find is the strength to stare straight at the despair, with our faces turned always to the Earth and never turning away.

We will bear witness, in every way we know how, on every sidewalk and every page, to the glories of this world and the sins against it.

We will understand that we are daughters of the Earth, pulled from her spinning surface. And so we will take on the duties of the moon. We will not look away. The shadow of the Earth will pass over our faces, but it will not erase us; at the edge of that moving shadow, our faces, our characters, will be most clearly seen. We will reflect the light that comes to us from the darkest spaces of the night.

This is what we were born to do. We swore it. There, at the wooden table, we swore it.

We put the food away and swept the floor. We gathered a bag of red apples and shook the trees; the deer who came in the night would find apples in the dew. The moon set in the orchid sky behind the hills, even as it rose on the other side of the Earth.

❖

WHAT DOES IT mean, this call to witness?

Let us be chroniclers of loss. Let no species disappear without public notice. If our ways of life are going to destroy infinitudes of lives, let us at least do it knowingly, and grieve for the terrible absence. Fill the forests with death notices. Transform every stump in the clearcut into a cross, so no one can drive by a bare-ass hillside without seeing it for what it is—a graveyard that stretches for miles. Let the roadsides bloom with shrines adorned with flowers to mark the extinctions of sparrows. Post missing-persons notices for the white-headed woodpeckers that used to frequent the ponderosa forest. Send an obituary to the newspaper each spring, when the frogs do not sing. Howl across the lake for the gray wolves that once roamed Oregon. Assemble the choir and sing hymns as the bulldozers gouge out the last checker lilies in the valley. Print pictures of ivory-billed woodpeckers on milk cartons and implore people to send any news. Rent a hearse and follow the truck that sprays poisons in the ditches. Have faith in human goodness; that if we could understand the enormity of the killings that are incidental to our casual decisions, we could find it in ourselves to change our ways.

Let us be chroniclers of danger. If there are warnings on cigarette packs, surely there should be the same warnings on gas pumps:

WARNING: Fossil fuels are addictive.

WARNING: Burning fossil fuels can harm your children.

WARNING: Burning fossil fuels can cause fatal lung disease and cancer.

WARNING: Burning fossil fuels during pregnancy can harm your baby, directly, through pollution, or indirectly, by damaging the planet's life-giving systems that will sustain that small person into the future.

WARNING: Burning fossil fuels can kill you and (so far, in tandem with habitat destruction) kill 40 percent of the plants and animals on Earth.

WARNING: Burning fossil fuels causes harm even to those who cannot afford to burn them.

WARNING: Quitting fossil fuels now greatly reduces serious risks to your health, the health of the planet, and the health of future generations.

Let us be chroniclers of wonder. Speak of celebration. How is it that humans sing in harmony, bringing different voices together into chords that shake the leaves on the trees? Where does dancing begin? Why does a child's laughter make a father cry with happiness? Can the universe celebrate its splendor without us, or does it need human joy? Speak of pollination. Speak of rain.

Speak of imagination, the biochemistry of new ideas, the frazzle-ended nerve cells reaching out to each other or the divine wind whispering in the soul. How can we think of what does not exist? How do we hope for what we have never seen, never in the history of the world? Speak of decomposition. How does it happen that death hurries back into life, even in the course of a season? Why is the skeleton of an alder leaf so beautiful when its life is gone and scattered? What would it sound like if we could hear all the chewing underfoot? Study the beginning and the end? How does it come to be that we are here, humans on the blue Earth, precariously perched in the middle of the perfect cosmic explosion? These are things we will never completely understand. But celebrating them makes all the difference in the world. The wonder of the ongoing processes of creation makes us humble. Their power scorches every bit of arrogance from our raised heads.

Let us be chroniclers of anger. It's wrong to refuse to hear. To close your eyes and plug your ears and make *blah blah* noises with your tongue is what kindergarten kids do, not senators, but at least the kids know it is obnoxious. There is an obligation to understand; it grows from the gift of our quick minds and knowing hearts.

It's wrong to silence another person. Jeers and insults are one thing. Threatening the tenure prospects or jobs of scientists to silence them is a violation of the right to free speech. Buying the silence of scientists by controlling the questions they ask and the articles they publish is a violation of the very essence of science itself.

It's wrong to keep silent when speaking out can prevent or reduce harm. Silence allows wrongdoing to continue unchecked. For god's sake, how many examples do we need? SEE SOMETHING. SAY SOMETHING. The poster at the security checkpoint at the airport is the same as the sign in the passenger train is the same as the anti-bullying poster in the schoolyard.

Here's what I have learned from experience: It's hard to talk about climate change. People don't want to hear it. They turn away—in guilt, in exasperation, in hopelessness, in fear, in despair for their children, in reluctance to make any changes in their lives, in embarrassment to see me acting like a kook, who knows? But for everyone who turns away, there is another who is relieved to finally be able to talk about what she has been holding in her heart, a secret that can finally be told. "Yes, I too am worried about climate change; what shall we do?"

invincible ignorance

I SWEAR TO GOD, it's happened again. I'm on book tour with *Moral Ground*, a call for moral action to avert the worst effects of a warming and degraded planet. The audience is convinced; climate change is here. They are empowered; nothing is stopping them from dramatically changing how they live and calling their leaders to account. And the first question out of the box is, "But what can we do about the people who deny climate change altogether? How do you change their minds?"

Full disclosure: What I say is, "I don't do anything about the deniers. I don't try to change their minds. Compassion and efficiency both would advise me to let them wise up on their own. People aren't irredeemably stupid, and time is a great teacher." But what I really want to say is, "I'm not worried about the deniers; I'm worried about the hypocrites, people like me (and you) who shake our heads at the dangers we face, truly worried, unable to sleep, but make few changes in our own lives and point our fingers at the deniers."

Okay, so let's take a minute to review the facts. "Scientific evidence for warming of the climate system is unequivocal," according to the Intergovernmental Panel on Climate Change (IPCC), a board of the world's leading climate scientists. Evidence is measurable.

NASA's list includes sea level rise, global temperature rise, warming oceans, shrinking ice sheets, declining arctic sea ice, glacial retreat, extreme weather events, ocean acidification, and decreased snow cover. We've heard this a thousand times: Ninety-seven percent of climate scientists agree that climate change is real, it is happening, and it is dangerous.

"But maybe climate change is not caused by human activities," deniers protest. "Doesn't matter," I say. "Either it is or it isn't. If climate change is caused by human activities, then we need to change the nature of human activities. And if climate change is *not* caused by human activities, then we need to change the nature of human activities, so that our own contribution to greenhouse gases doesn't make the effect of the natural causes worse. Either way, we've got to act."

Say your child comes to you sweating, with a terrible fever. Say you didn't cause it. Do you wrap her in steaming blankets that will make her fever worse? Do you put hot towels on her head? Do you light fires around her? You do not. Even if you didn't cause the fever, you refrain from doing what might make her hotter still. Knowing how dangerous a fever can be, you do everything you can to bring her fever down.

But, oh, how tired I am of these endless, fruitless—and most importantly, *useless*—arguments. In 2014, the Yale Project on Climate Change Communication asked Americans to respond to this question: "Do you think climate change is happening?" The answer? Yes: 64 percent. No: 19 percent. Sixty-four percent of the American populace is plenty enough to push action to reduce greenhouse gases—if we all tried. If we wait until everybody is on board, this train is never going to leave the station.

In a Fox News poll, 8 percent of Americans said they are sure that Elvis Presley is still alive and 11 percent are undecided. But undertakers didn't wait to bury him until every single American agreed that he was dead. They put his poor bloated corpse in the ground. Same goes for climate change deniers. Let's move on.

But there's another reason that I don't waste time trying to convince people that climate change is real. It has to do with the principle of falsifiability. This is a principle that allows one to distinguish between a statement that is empirical (based on evidence)—as all scientific claims must be—and one that is something other (an expression of faith or belief). A claim is empirical if it is possible to imagine a fact that could show it is false. Suppose I say, "Elvis is still alive." If I admit that it's possible that some evidence would show I was wrong—exhumed bones, for example—I've made a (false) empirical claim. However, suppose I say, "Elvis is still alive, and there is no possible evidence that could count against the claim." Exhume the bones from his grave? Somebody else was buried there. Find a death certificate? The doctor is in on the plot. Because no evidence could falsify the claim, it is not an empirical claim but a statement of belief.

So what of the claim that climate change is a hoax? Arctic ice is melting? Must be solar flares. Acidity of the ocean is rising? Must be underwater volcanic activity. The temperature of the Earth is increasing? Impossible to measure. If there is no possible evidence that could count against the hoax claim, then it's a statement of belief, not a claim of fact, and I have no interest in arguing.

Old logic textbooks call the refusal to revise one's views in the face of contrary evidence the Fallacy of Invincible Ignorance. This decoupling of evidence and belief means that denial might best be understood as an act of will, or loyalty, or economic self-interest, or political strategy. Is there any logic to this denial? Actually, there is.

Philosopher Michael Nelson and I think we have the logic figured out. Here is the article we wrote together. It starts with the same practical syllogism we have been working with all along:

THE LOGIC OF DENIAL

Consider the logic by which people reach policy decisions of all types. Any argument reaching a conclusion about what we ought to do will have two premises. The first premise lays out the facts of the matter, delivered by the very institution that society has charged with providing us facts: science. With regard to climate change, the facts are clear: Unchecked anthropogenic climate change will profoundly harm the chances of the world's children and future generations, undermining the necessary conditions for human thriving. The second premise lays out the values at stake, a culture's collective moral wisdom about what is just and good. These are equally clear. It's wrong to hurt children. It's a massive violation of human rights to condemn multitudes of people to struggle and misery. When you combine these facts and these values, the conclusion is inescapable: We are obligated to act quickly and dramatically to avert anthropogenic climate change.

If deniers want to reject the conclusion of a valid argument—which is exactly what they want to do—they have only two strategies. They could, of course, shrug off the moral principles. "Hurting children? Fine with me." "Violating basic human rights of billions of people, present and future? No problem." But no one would use this strategy; that would reveal a moral failing or sociopathology of breathtaking proportions.

What's left? The only alternative is to deny the facts of the matter, undermining or profoundly misunderstanding the scientific method and motives of scientists. To endlessly, mindlessly quibble over the reality of melting sea ice only makes you, at worse, stubborn or stupid. But to quibble over whether we have a moral obligation to protect children from harm makes you dangerously immoral. It's an easy strategy decision: Go after the facts. Thus millions of dollars have been poured into attacks on climate science and scientists—dollars supplied by those deeply invested in preventing society from drawing any conclusions that might block the unimaginably profitable activity of pouring carbon into the air.

We can learn from this. First, we should not write off climate change denial as yet another example of scientific illiteracy, or further evidence of declining faith in science as a source of explanation, or as a lack of communication prowess among scientists. That's not what's going on here. Those who do not want to take action against climate change, for whatever reason, find it easier to undercut the science than to engage in real dialogue about the values. So we have a national climate change debate that is marked by a furious, often fallacious, certainly futile debate about facts. But the national discourse about values—the conversation about what we most deeply value in our lives, about what we most owe the future—has gone missing.

Second, we should realize there's no point in debating the science. It's likely there is no science, no level of certainty or consensus that will change the denier's mind. That's a smoke screen, a black hole of effort to keep the rest of us busy, to push off the implications of the argument one more time. The deniers want to reject the conclusion of any argument for meaningful climate action, and their professed rejection of the science is merely a means to that end.

There may be a kind of delusion here, though perhaps not the kind we think. Surely, there are many hapless people deluded by attacks on climate science. But those who launch the attacks are not deceiving themselves about the facts; they know better. For them—the organizers of the campaign against meaningful climate action—climate change denial is not a matter of ignorance or mistake or delusion but a strategic decision. What they really must believe, but cannot say, is that greed or self-interest or limitless profit trumps the safety and happiness of children and the human rights of present and future generations.

These are the beliefs requiring a full-blown public debate. Do we have obligations to future generations? Do we have obligations to rescue children in danger? Do we have an obligation to respect human rights? And above all, what are the limits to the values we would sacrifice and the moral principles we would violate to make a killing on investments in gas and oil?

❖

AND NOW I get a Christmas card from a high school friend. The highlight of his family's year was the purchase of a Buick Enclave (an SUV that weighs five thousand pounds, named after exclusiveness itself), on which they had already put twelve thousand joyous miles. "Happy New Year," I responded. "My work next year will be on climate change." "Terrible problem," he wrote back. "What really angers me are the people who deny that climate change exists." Fine, dear Bobby. But what really angers me are people like you, who know better.

I didn't write back. Sometimes I am so snarky that I'm ashamed, and in high school, he was always smarter than I was anyway.

As I say, I'm not much worried about the *old* climate deniers. On this one, they're on the wrong side of history and all the evidence of science. But my former student, Mary DeMocker, who blogs as Climate Mom, warns about a new species of denier. "I don't mean global warming deniers. No, New Deniers are scarier. They deny that action can help. Here's their litany, reduced: Fossil fuel runs everything Citizens United killed democracy everyone's busy youth are zombies we're frogs in a heating pot it's too late and you'll never outspend Big Oil."

It's dispiriting and downright dangerous to engage in debate with a New Denier—sort of like trying to rescue a drowning man who is flailing away, trying to climb onto your head. So what does Mary say to those who tell her she can't save the world? "'Watch me try, suckas!' We're alive at THE most extraordinary point in human history. Climatologists say we can avoid catastrophe by cutting global emissions 6 percent yearly—starting today. . . . So help me or Get. Out. Of. My. Way."

false promises and dead ends

REDUCING THE RATE of global warming is going to be really, really hard. Plans to take the necessary steps will face powerful opposition. So who can blame people who turn to related tasks that have perhaps a greater chance of success?

It makes a certain sense. Searching for an analogy leads me to John Muir's little dog Stickeen, who had followed the old adventurer-philosopher onto a glacier. Late in the day, the dog found himself separated from safety and from his beloved John by a long, deep crevasse that could be crossed only on a treacherous ice bridge. The ice bridge was narrow. And slippery. Absolutely, it was dangerous. What was the little dog to do? Obviously, the most direct course would be to screw up his courage and cross the bridge, as carefully and as safely as he could. But is that what a frightened dog would do first? Maybe he would wander around for a while, thinking he could figure out how to live just fine on the ice without his master. Maybe he would howl in fury at anything handy—the wind, the crevasse. Maybe he would take the time to build up his snow-bridge-crossing skills, practicing putting one foot in front of another. Maybe he would lie down in the snow and die of cold and despair.

Obviously, the most direct course in a time of climate change is to do what it takes to get from a fossil fuel economy to an economy based on thrift, sharing, and renewable energy sources. But we shiver and dither. Some of us try to figure out how we might live on a hot planet awash in sour seas—call it adaptation. Some of us turn impotent with rage, lashing out at Democrats, windmills, enviro-wackos, and a bunch of others not at fault. Call that scapegoating. Others will try to build our skill sets and reorganize our lives to lessen the harm that climate change will cause us. Call that resilience. Some of us will give up in despair: "If nothing I can do will make a difference, I will do nothing."

False promises, all of these, and dead ends that keep us from doing the work of slowing climate change before it's too late.

ADAPTATION

As the fury of climate change kicks in, I can feel the heat on the wind, smell the sting of smoke from wildfires. I believe I also smell in the air a significant social change—a dramatic increase in planning and funding for adaptation to global warming, and a corresponding decrease in attention to stopping the greenhouse gas pollution that causes it. New York proposes a $20 billion project of flood walls, levees, and bulkheads to protect the city from storms. Rotterdam announces its "Rotterdam Climate Proof" plan to make the city "fully" resilient to climate change impacts by 2025, a plan that includes floating bubble pavilions. With the same boosterish spirit, more than a hundred U.S. cities now have climate adaptation plans in place—and who can count the special commissions, committees, agency teams, and panels of experts charged with figuring out how their own citizens can continue to live as they always have, while seas rise, storms escalate, forests burn, crops fail, and in all its devastating ways, carbon catastrophe grips the planet by the throat.

The driving assumption is that humans *can* adapt, continuing to thrive on a sizzling, stripped-down, drought- and flood-stricken, dangerously destabilized planet. That assumption is shaky, given that unpredictable conditions might be the only contingencies one can't plan for. But leave that aside. Leave aside also the frustrating question of why anyone imagines it will be easier and cheaper to adapt to a devastated world than to mount a full-out effort to slow the wreckage and save what's left of the life-sustaining ecosystems. To my mind, the worst part of all this is that the very moral failings that characterize climate change itself are being replicated and amplified in many of the plans to adapt to it—as if storm and extinction had taught us nothing about justice or reverence for life.

Let us begin to address this issue, as professors sometimes perversely do, with a multiple-choice quiz:

When your house is on fire, what should you do?

a. Not one damn thing.

b. Defame the people who called 911 to report the blaze.

c. Debate whether the fire was caused by humans or natural fluctuations in temperature.

d. Write a grant to study the effects of fire on two-by-fours.

e. Formulate a business plan to corner the market on corrugated metal roofing for hovels.

f. Appoint a commission to study how to adapt to life in the burnt-out husk of a house.

g. Put out the damn fire.

Hint: Imagine that your children are in this house, and not only your children but about 1.9 billion other children. Imagine that this house is beautiful beyond imagining, that you have been happy in this house, a sheltering, nourishing place that provides water and warmth and food. Imagine, in fact, that this house is the only possible source of everything your life and happiness depend on. Imagine that there is still a chance to save the house, or at least large parts of it. But it's a narrow, perilously narrow, chance—and it depends on throwing everything the world has at the fire.

Now what shall be said about all the agencies, commissions, and special committees that are laying down the fire hoses, pulling back the ladder trucks, pouring coffee all around, and devising elaborate plans for how humans might be able to continue to thrive in the smoldering remains? How shall we politely respond to Rex Tillerson, chairman, president, and chief executive officer of ExxonMobil, who says, "As a species . . . we have spent our entire existence adapting. So we will adapt to this"? "This," of course, refers to global warming.

Over the past decade, the United States has worked its way with excruciating slowness through steps (a) through (e). First, polite people didn't talk about climate change (a). Then the fossil fuel industry let the dogs out to chew on scientists and shred their homework (b). Once climate change became undeniable, the industry delayed every action with debates about its causes (c). Scientists, policing themselves against advocacy, studied the unfolding cataclysm and drew every conclusion except "This is madness" (d). Capitalists smelled opportunity on the winds off melting ice sheets, where there were new shipping lanes and new oil to be exploited (e). And here we are at option (f), systematically trying to figure out how to adapt to life in the burnt-out shell of what was once the most beautiful and nourishing homeland in the solar system, when what of course we should do is put out the fire (g).

Adaptation: adjustment to environmental conditions. *Accommodation* might be a better word, or *appeasement*—learning to live with the destruction rather than mustering forces to stop it. The fossil fuel industry must be gleeful about all the adaptation proposals; it's the final victory of profit over sanity, convincing people that they should pay to protect themselves from climate change, even as both climate change and oil industry profits grow unchecked. Our work right now is to stop greenhouse gas pollution, not figure out how to live with its terrible costs. We can't stop the climate change that fossil fuels have already set in motion. But we can stop making it worse, and then we can know that we did our best to leave the next generation a world from which they might salvage something nourishing and beautiful.

I understand that it's prudent to figure out how to live in the world as it presents itself. Let that be clearly said. But a single-minded focus on accommodation to climate change, a focus that ignores the need to reduce climate change, is a moral failure. That's my claim; here are my reasons.

On a purely consequentialist calculation, the opportunity costs of accommodation prevent it from penciling out. "An ounce of prevention is worth a pound of cure" is especially true when what is being prevented is irredeemable damage to the natural systems that support all life. The danger is that "adaptation" can become a smoke screen that hides or minimizes the real consequences of global warming and so delays action, which allows people to continue to live as they always have and the fossil fuel industry to continue to profit as it always has, while the world misses its last chance to stop runaway climate change. Adaptation is a false hope, in a time when all of us, myself included, are tempted to cling to any hope, false or not. But if it delays or derails action to stop fossil fuel pollution, adaptation will not create the greatest good for the greatest number, not by any calculation.

What if the money New York City proposes to spend to "adapt" to climate change were spent instead on stopping its advance? Twenty billion for a massive reforestation of the planet. Or twenty billion for local solar projects. Twenty billion for a new agriculture that keeps the carbon in the soil. Thousands of U.S. towns have funded commissions on adapting to climate change. How many have commissions on stopping the pollution that causes it, with subcommittees on badgering Congress? Task forces on empowering officials who have integrity and good ideas? Subcommittees on hounding out of office those who, disguised as elected officials, are paid lobbyists for Big Oil? Blue-ribbon panels to collect from the fossil fuel industry the costs they have externalized onto New York City and the world?

But a focus on adaptation is not just imprudent. It's unjust. It perpetuates and then magnifies the basic injustice of climate change: that the people who benefit from the profligate use of fossil fuels, the people who are causing global warming, are not the ones who are bearing the costs. With "adaptation" projects, the privileged can use their power and money to try to shield themselves from the worst consequences of their own excess while imposing the costs of climate change on the disenfranchised and displaced—and imposing not only the cost of the profligate use of fossil fuels but the additional costs of our hapless efforts to adapt to the global warming that results.

Accommodation is a further injustice to future generations to the extent that the new infrastructure of accommodation increases the greenhouse gas load in the atmosphere. Seawalls and desalination plants and dams and drainage pipes are made of concrete; it takes four hundred pounds of carbon dioxide to make every cubic yard of concrete poured. The bulldozers shoving around sand dunes gulp diesel and burp smoke. The projects might buy a little time to eke out ways to keep doing what we are doing, but only by creating additional climate change costs that we will pass on to our children, at compounding rates of interest. It is morally unjustifiable to take from the world whatever resources we need to protect ourselves from our ongoing recklessness and send the bill to the future in order to fuel the greatest going-out-of-business sale the world has ever seen.

Accommodation is unjust also to the poor and dispossessed. When Rex Tillerson says we can adapt to climate change, who is this "we"? Is it African children on failing farms? Is it northern people on melting ice? It is coastal residents of Bangladesh? Or is it Rex Tillerson, who earned $40.3 million last year? Ambitious plans for accommodation allow privileged people to continue burning oil and gas, on the probably illusory faith that they are going to be all right. It's bad enough that fossil fuel bonfires are destroying the material basis of the cultures of people in Africa, the Indian subcontinent, low-lying Asia, the Arctic. But what truly moral society responds to the suffering it has caused other cultures by investing billions of dollars to make sure that the same thing doesn't happen to its own culture?

And this says nothing of the cost to plants and animals. Adaptation means that individual plants and animals will die, and that has to be accounted for. But adaptation also means that whole species and ecosystems will perish: the great outpouring of life that has evolved in a stable climate over millions of years, the silver birch forests and taiga plains with all their small, shimmering lives, the dawn chorus of birds and frogs singing around the globe, the nest of shivering hares, plants and animals that are startling, and maybe even miraculous, in their adaptations to this world—*this world*, not the world fundamentally changed.

"There will be losers and there will be winners in the adaptation game," biologists assure the public. But it's the losers that call to the conscience. And it's not a game.

Plant and animal species can adapt, it's true. The mechanism of adaptation is dying. Dying in huge numbers. Unless the species goes extinct, the more individuals die, the faster the species adapts. Adaptation demands starving nestlings and kits drowned in the den and the seared roots of seedlings. Studies of salmon and Darwin's finches show that it takes about 90 percent mortality to achieve rapid adaptation. When this is caused by human choices, it is a morally unacceptable cost to the creatures of the planet, who did not bring this calamity on themselves—a price exacted on other beings by humankind's failure to count other lives in the narrow calculation of their self-interest.

I believe that humanity is called to a kind of adaptation. But not adaptations that harden present human patterns and ideas, reinforcing with steel beams the patterns of excess and exploitation. Rather, we are called toward creative change in the very ideas of what it means to be a human being and how we might once again blend humanity's moral imagination into the creativity of the unfurling universe. That is the true adaptive challenge.

In June, I went with Frank to a cabin on the east fork of the Toklat River in Denali National Park to think about global warming. It didn't go well. Approaching midnight in June, the thermometer in the shade read ninety-three degrees. The sun, anchored in the northwest sky, fired rounds of heat against the cabin's wall, even though we had hung a blanket from the porch rafters to shade the door. I was lying naked on the bunk, slapping mosquitoes. Next to the wall, my husband lay completely covered with a white sheet, as still and dismayed as a corpse. Frank would rather be hot than bitten, and I would rather be bitten than hot. Neither of us would sleep, even if we could, because we had left the door open to catch each vagrant breeze. That meant that nothing but a screen stood between us and any grizzly who wandered by. My assignment, being closest to the hypothetical bear, was to be ready at any moment to leap out of bed, slam shut the heavy wooden door, and slide home the bar.

This was record-breaking heat, a spike on the graph of jaggedly rising temperatures in Alaska. The average day in Alaska is now four degrees Fahrenheit warmer than just a few decades ago, and seven degrees warmer in the winter. The Arctic is heating twice as fast as the rest of the world. Researchers at Denali National Park are developing protocols for monitoring long- and short-term climate trends, but I didn't feel a need for any numbers. I could smell the heat on the wind, the sting of smoke from wildfires. I could see the warmth on the mountains, bare of snow in June. I could hear the lichens crackle under my feet, and between my teeth, I could taste the dust that blew in whirlwinds off the riverbed.

This measuring and monitoring—here and around the world—was not aimed at stopping the greenhouse gas pollution of the atmosphere. Rather it was laying the groundwork for decisions that managers of all sorts are preparing to make about how civilizations will adapt to whatever global climate change throws at us.

Lying naked and nervous on my bunk, I mulled this over. Across the river in front of our cabin, Dall sheep grazed high on the sides of the mountain. Up there, among the rocks and the clear expanse of space, there's good, nourishing grass and heather in full-bell bloom, and willows no higher than the sheep's withers. Yesterday morning, we caught our breath just looking at them as they grazed across slopes so steep that a stone they dislodged bounced a half mile before it found repose, slopes so high that they poke into the sky. Nothing blocks the view from up there. That's a good thing: Sheep can see wolves coming from a long distance away.

But spruce forests are growing higher up the mountainsides into the refugia of the Dall sheep. This new atmosphere, so rich in carbon dioxide, this new weather, so mild in winter, gives spruce trees and shrubs everything they need to grow on land that had been denied them. The expanding range is visible already on the slopes, with new trees higher on the mountain every spring. The sheep will be forced higher on the peaks, away from the shadows that hide wolves. Fear will keep them moving into the last tufts of grass.

What will the sheep do then? I can imagine them going higher and higher, into air thinner and thinner, squeezed tighter and tighter by the wolves and the trees. A wolf lunges from the cover of the forest and the last ewe leaps away. But there are no mountaintops left to leap to, and she is suspended in space like Wile E. Coyote, moving her legs gracefully, sailing out of Earth's orbit and on and on, running or swimming, you can't tell, until she circles a new planet and descends slowly to land with a short hop on the extraterrestrial mountaintop. Because that's what people say, isn't it? That when we use up this planet, we can go to another, and maybe the sheep hope there will be mountain avens blooming in the new place, and dwarf willows on the fellfields.

We were up early yesterday, driving through the high tundra on Thorofare Pass, when a snowshoe hare jogged across the road. It saw us and froze, as hares will do, confident that its camouflage hid it in the birch and shadows. But it would have been appalled if it had looked in a mirror. Blotched with white, that little animal flashed like an airport beacon.

The problem is that snowshoe hares turn white in the winter, brown in the summer; the change is triggered by the length of the days. And that worked when snow disappeared reliably as the days lengthened. But snow melts in response to temperature, and with the early warming of the Arctic, snow gives way to soil before the hares turn from white to brown.

What will the hares do? Adapt, most surely. Which, given the harsh rules of natural selection, means they will selectively die. If there are hares that by some chance change color earlier in the spring, they may live. The rest will not. Or maybe the hares will get an idea for a disguise—they can lick themselves all over and roll in the dust, for example. Or maybe they'll find a technological solution, a hair dye made from bark or some such. But maybe they won't. I don't doubt that there will be winners and losers in this global warming adaptation game. A white hare hiding in a brown thicket is most certainly a loser.

Meanwhile, the scientists are on their knees in the sedge meadows at Eight Mile Lake, poking hypodermic needles into the dirt. What they are learning makes my heart sink. Much of the northlands rest on permafrost, a layer of ice and frozen organic matter that holds carbon—almost half the Earth's carbon, taken together. When the permafrost melts, as it is doing at varying rates of terrifying, the carbon-bearing matter decomposes and releases carbon into the air. At first, this is a wash, because the carbon dioxide encourages the growth of plants, which sequester carbon. But soon enough, the plants can't use all the carbon dioxide. Increased CO_2 traps heat, which creates more melting, which releases more CO_2, which traps more heat. This is the feedback loop that can spiral out of control—as if it has ever been possible for humans to "control" any global system. Disrupt, yes. Control, not so much.

Melting permafrost also releases massive amounts of methane and nitrogen oxide, which trap three hundred times more heat than carbon dioxide. Already, every day, the greenhouse gases in the atmosphere are trapping excess heat that equals the heat of four hundred thousand Hiroshima atomic bombs. I'm no mathematician, but I know how to multiply. And suddenly this isn't just about Dall sheep or snowshoe hares. It's not just about the Arctic.

We will adapt to this, you say, Rex? Maybe. But what will our lives be then?

Sure, we can get along for a while. We can air-condition Alaskan bush cabins and ship in deadlier mosquito poison and start an adopt-a-Dall-sheep program. But note the futility of this: Each of these efforts to adapt puts more carbon dioxide into the atmosphere and kicks the climate a bit more out of its track. If adaptation requires new products and machines to transport them, and new products and machines require resource extraction, and resource extraction requires fossil fuel energy, I don't see that we've made much progress. We have merely succeeded in foisting onto future generations not only the costs of burning fossil fuels but also the additional costs of our hapless efforts to adapt to the global warming that results. In fact, the illusion of adaptation is starting to look like a great way to increase the market for oil. That is a seriously bad joke on the world.

I would be calmer about all this if I believed that global warming were taking us to what people call "the new normal," a fixed world, hotter than this but still something to adapt to. But that's not how it works. If we don't stop it, climate change will be an ongoing spiral of chaotic, unpredictable change: wobbling ocean currents, erratic jet streams, violent weather, falling off tipping points, kick-starting new feedback loops, sharp, violent leaps. One of the most dangerous of the chaotic, unpredictable changes is the increased chance of war, I fear. The list of the causes of war reads like the list of the consequences of climate disruption: economic collapse, threats to essential resources, existential threats such as starvation and loss of livelihood, massive movements of refugees. Both history and philosopher Thomas Hobbes argue that people can adapt to lives of constant warfare, but, in Hobbes's words, what "solitary, poor, nasty, brutish and short" lives those are.

And honestly, I would be calmer about the push for adaptation if I had much faith in technology or the human ability to deploy it wisely. We're used to technological substitutes for the natural systems we destroy. Ruin a river? So build a fish hatchery. Drain an aquifer? So build a desalination plant. Starve people off their land? So build a city of hovels. But the sorry lesson of history is that technological substitutes, designed to solve one problem, almost always create new problems. Another sorry lesson is that new technologies seem to have instincts for self-preservation as strong as any animal. Or should I say that each technology serves a subgroup of investors who strive mightily to keep it churning out profits, even when the terrible cost to other people and to the commons becomes clear? Technological innovation, if it is built to the specifications of the same ruinous spiral, will not save us.

For a time, until global warming engulfs us all, the privileged will use their money to shield themselves from the most awful consequences of their own excess. Those without that privilege will die, or see their children die, or watch their cultures and governments collapse. And so the illusion of adaptation empowers the continuing exercise of a calamitous selfishness.

I consider myself pretty adaptive. But I can't adapt to global injustice. I can't adapt to betraying my grandchildren. I can't adapt to a world without songbirds and all the other lyrical little lives. And I can't imagine adapting to any of this to increase the profits of Big Oil.

It's true that the greenhouse gas pollution already in the atmosphere is going to cause climate changes that will work themselves out over millennia. Given that, what we have to do is the one thing we can do: We have to stop making it worse. Which means we have to stop releasing greenhouse gases. We have to leave the coal, the oil, the natural gas in the ground.

SCAPEGOATING

Our boots kicked snow into the beams of our headlamps as we hiked up the trail. Lights shot against the sparkling trunks of ancient trees, glanced toward the trees' invisible crowns, crisscrossed over the mounded snow of the trail ahead. As the trail grew steeper, the lights gleamed in clouds of frozen breath. "We're getting close. Better turn your headlamps off," our guide whispered. The students did as they were told. These were students in my field course, who had come to the forest to think about problems in environmental ethics. They were about to encounter a doozy.

The moment when everyone turned their headlamps off was intensely dark. We moved forward more slowly, guided by a vague light that seemed to come from the snowy trail itself. No one said a word; some held hands, mitten to mitten. The forest to either side was utterly opaque. At a place where the trail widened into a clearing, the guide held up his hand. We settled into silence. At first, there was the rasp of nylon parkas, then nothing. A sift of snow falling. A limb breaking under the load of snow.

Whispered instructions: "We'll hoot for the spotted owls first. If we gave a barred owl call first, the spotted owls would never dare to call." We knew this, that the little spotted owls had good reason to be afraid of the barred owls that were moving into their dark forests.

"Like this. Five hoots." We all jumped as the guide hupped into the night. "*Hup . . . hoo-hoo . . . hoo-woo. Hup . . . hoo-hoo . . . hoo-woo.*" Students giggled, then hushed themselves.

"Try it," the guide urged. A bold student hooted and fell into silence. Another called. Another. Then they stopped calling to listen. In a silence that deep and long, you can hear the snow fall.

"One more time." This time, it was a conversation of student-owls as enthusiastic as Friday night at the sports bar, and a silence that turned, after a long, long time, into a murmur of regret. Students pulled off their woolen hats to hear more clearly, turned their heads. Nothing but snow falling in the dark. The guide screamed—there is no other way to describe it. "Sometimes this encourages them," he said, but the students were too startled to do anything but laugh. "Shhh." They shushed. Darkness. Snow falling.

"Okay, let's try the barred owls. It's *Who cooks for you? Who cooks for you-all.*" He demonstrated, and this time the students were more nuanced in their calls. But right away: "Shhh. There it is." They listened intently. A student pointed up the hill to the west. Yes, vaguely, muffled: *hoo hoo huh HOO*. Students rustled with excitement. After a long pause, another call from the guide, and another muffled response, closer this time. A few students called. Snow fell and fell, covering their hats and shoulders with snow, but there was no answer. The guide hooted. Nothing more but that dark night, when the students knew that a barred owl was drifting silently through the forest, and maybe—maybe—a spotted owl was cowering in a rotted spar, silently listening to the snow and the faraway voice of its bigger cousin.

The U.S. Fish and Wildlife Service had just released its plan to shoot barred owls. For centuries, the barred owl was an eastern bird, living in the edges of cut forests and clearings. But it had recently crossed the continent, hopscotching from one clearcut boreal forest to the next, and moved down the Pacific coast into the ancient forest range of the smaller spotted owl, disrupting their nesting, outcompeting them for food, and ultimately driving them off. The northern spotted owl is listed as an endangered species.

It's apparently not hard to shoot a barred owl. Graduate students—"graduate research assistants"—can be hired to do it. Just call the owl in, aim, and shoot. In one of a series of experiments, the Fish and Wildlife Service was planning to kill up to thirty-six hundred barred owls in California, Oregon, and Washington. If the experiment is a success—if spotted owls return—the Fish and Wildlife people will kill more barred owls in other places. I wanted my students to think hard about this. It's tough. The more that human decisions disrupt the balance of ecosystems, and the more human decisions bring animals to the edge of extinction, the more desperate and costly become the efforts to save at least a remnant. Desperate circumstances, desperate measures.

So conservationists end up killing animals to save animals. Some of these strategies are successful and relatively uncontroversial. Spraying tons of poison pellets from helicopters, conservationists eradicated ship rats and Pacific rats from four Maori-owned islands south of New Zealand, protecting populations of sooty shearwater and other species. In more controversial actions, wildlife managers "used lethal measures to remove" (shot) California sea lions at the Bonneville Dam on the Columbia River, to reduce losses of salmon and steelhead. Decisions that sacrifice one species for the sake of another test the limits of usual moral decision-making tools.

The temperature warmed up that night and rain fell on the snowdrifts. By morning, the sun was out and there was a hard crust of ice on the snow. Carrying steaming cups of coffee and cocoa, students slithered to the library, often breaking through the crust and sinking a leg to the knee, leaving a splash of coffee on the snow. Students built a fire in the stone fireplace, and soon the room was warm with the smell of wet wool and chocolate. We talked about the owls we heard, and did not hear, in the night. No one felt it was right to kill the barred owls; the news articles we read made it clear that even the marksmen were in agony. But how can one bring reason to a dilemma so fraught with emotion?

I asked my students to think about the decision to kill barred owls, first using the usual approach to decisions about conservation policies—consequentialism, the moral theory that judges a policy right or wrong by predicting its consequences. Killing members of one species to save another is judged to be the right policy if the killing creates the greatest possible balance of desirable over undesirable consequences. In this context, scientific data about the effect of the killing on the values at stake are essential. They are not decisive, however, the students soon realized, because they do not decide which set of values is paramount when interests come into conflict: Should the decision enhance the welfare of individual animals, the preservation of a species, the thriving of an ecosystem, or—don't forget this one—human economic interests?

Society has to figure out what it values most, and so far it hasn't achieved anything close to an informal civic consensus. When managers employ only consequentialist modes of decision-making, they back themselves into an unnecessarily small moral box. In a situation where there are competing interests served by different strategies, every problem is a dilemma, forcing a hard choice between the interests of some and the interests of others—if you have only a consequentialist decision-making strategy.

However, consequentialism is only one of several moral decision-making tools available. My students have studied the wide sphere of ethical discourse. They can bring any number of moral theories to the problem of killing one animal to save another. Readers of this book have seen these before. Virtue ethics judges the morality of the act by asking whether it grows from traits considered virtuous: Are the motivations for killing barred owls, for example, honest and honorable, or tainted by self-interest or politics? Duty-based ethics judges the morality of the act by whether it conforms to duties, usually duties of justice: Is killing barred owls just? Are they the wrongdoers? Consequentialist ethics, as has been said, judges the morality of the act by the desirability of its expected consequences: Do the values gained by saving spotted owls outweigh the values lost by killing barred owls? Chances of making a well-considered decision can increase as policy makers bring all these tools into play, especially when the right course of action is not intuitively obvious.

Fortunately (or unfortunately), humankind has had a long time to think about the moral justification for killing some to save others. "What is the human institution in which humans kill some to save others?" I ask my students. The context, of course, is war, and from a long discourse about ethics in war comes the intellectual tradition known as Just War Theory. From Saint Augustine in the fourth century to Michael Walzer in the twentieth, philosophers have asked, "Given that it is prima facie wrong for humans to kill other humans, under what conditions is it justifiable?" By analogy, if one assumes that it is prima facie wrong for humans to deliberately kill members of other species, under what conditions can it be morally justified?

Just War Theory is a multivalued approach, using virtue, duty-based, and consequentialist tools. Consequentialist thinking is part of the deliberation, but it is constrained and bounded by justice and virtue considerations. The deliberation begins with questions of justice.

1. *Just cause.* What is the wrong or harm done that would justify violence in return?

The premise here is a principle of retributive justice—that only a substantial harm can warrant the use of the violence of war, and the violence must be directed against the one who does harm. The analogous issues in the case of killing for conservation purposes are these: Is the harm done to a given species substantial, and are members of the targeted species responsible for that harm? This condition requires careful science to establish a causal relation between depredation and decline. This research may be relatively straightforward, in the case of sea lions and salmon, or rats and shearwaters, but somewhat more difficult in the case of barred owls and spotted owls, and perhaps impossible in other cases in which many factors are in play. But without clear evidence of harm and causal responsibility, the first condition for justifiable killing is not met.

2. *Last resort.* Once a harm has been identified, have all nonviolent methods to rectify the harm been considered, tried, and exhausted?

On this principle, decision-makers should ask, "Is killing members of one species the only remaining way to improve the prospects of another?" The answer might be yes, as perhaps in the case of the rats on the Maori-owned islands. But other instances are not so clear. In the case of the sea lions in the Columbia River, for example, managers have indeed tried many alternatives. But if there remain nonviolent options that have not been tried—say, removing the river-blocking presence of the Bonneville Dam, most surely a cause of the salmon's vulnerability to sea lions—then the killing cannot be justified.

But suppose, for the purposes of discussion, that these two justice conditions have been met. Then the questions turn to a consideration of consequences.

3. *Reasonable prospect of success.* What are the chances that the killing will be successful in removing the source of the harm?

Deaths and injury incurred in a hopeless cause cannot be justified. This condition raises the "cannon fodder" question. Generals can send waves of soldiers into the line of fire, but will it win the war? How many sea lions will be killed before salmon populations recover? How many generations of graduate students will be sent into the forest with shotguns to save the spotted owl? The ultimate goal of a just war is to reestablish peace, a peace preferable to the peace that would prevail if the war had not been fought. Will an ecosystem be healthier, more robust, more resilient, after the targeted species is removed? Or will removing the targeted species create a fragility that requires constant intervention to maintain? If the strategy kills animals in a campaign that is doomed to fail, it cannot be justified. The challenge to science is to predict the chances of success, and the burden of proof is on those who advocate the killing.

4. *Proportionality.* Do the benefits of the killing outweigh the harms?

This is not a straightforward utility calculation of the sort familiar to conservation policy. For one thing, the question cannot be asked until the justice conditions have been satisfied. Secondly, the comparison is more properly between speculative or possible benefits (the spotted owl might be saved, for example), and certain and definite harms (a thousand barred owls will be killed). The scientific task is to envision anticipated and unanticipated outcomes. However, the uncertainty of science and the complexity of ecosystems make the calculation of benefits and harms even harder. It is difficult to be sure, in advance of the conservation effort, that the new situation will be preferable. In Just War Theory, that is reason to pull back and reconsider the strategy.

After these consequentialist questions, the questions address both virtue and duty-based considerations:

5. *Right intention.* What are the intentions behind the use of violence?

This question focuses on the character of those who kill and those who make the decision to kill. Saint Augustine called for a "mournful" disposition characterized by anguish and regret—a moral outlook that honors the dignity and worth of the foe. Bringing the right intention to the killing helps restrain the violence from becoming indiscriminate and disproportionate, as required by the next, and last, question.

6. *Discrimination.* Is the violence limited and restrained, and are only those who pose a lethal threat targeted?

In Just War Theory, the object of force is not first to kill or injure but to incapacitate or restrain. Even if this is not possible, a just war requires that the weapons must discriminate between combatants and noncombatants. The entanglement of soldiers and civilians is a moral challenge for those who plan wars, and species are no less entangled. What other populations are indirectly harmed when a given species is removed from a system? Here again, deep knowledge of ecosystems is essential to making a justifiable decision about killing a targeted species.

This principle invokes the retributive-justice prohibition against harming the innocent: No person should be punished for the wrongdoing of another. The moral danger is that a conservation-by-killing strategy might end up harming one species when it is another species that is to blame for the damage. This is scapegoating, a biblical reference to the practice of offloading the sins of the people onto a goat. (Leviticus 16:22: *and the goat shall bear upon him all their iniquities unto a solitary land.*)

Are sea lions responsible for declining salmon populations, or are sea lions caught in the crossfire of human practices that close rivers to salmon? Are barred owls to blame for reducing populations of spotted owls, or are they finishing up a job that began with the cutting of boreal forests and coastal old growth? The risk warned against is that targeting an innocent species will make it harder to envision strategies that could make a real difference, such as changing the human and corporate practices that caused the harm in the first place.

HEAVY WITH ICE, Douglas fir branches fell in the forest, and a red cedar cleaved off a branch that was itself the size of a tree.

In the library, we wrestled with the dilemma of owl v. owl. Very few dilemmas are clear cases of this or that, black or white, barred owl or spotted owl, owls or jobs, climate change or nuclear power, despair or hope, on and on. "Always be on the lookout for false dichotomies," I suggested to the students, "especially when a dilemma offers a choice between two nasty alternatives and forces you to do what you think is wrong to avoid a greater evil. Ask a few questions: Whose interest is served by presenting a problem as the choice between two stark alternatives? What caused our choices to become so limited? What is the third way?"

I think about the thirst of that suffering biblical goat and the relief of the villagers, finally to be rid of it and the stain of their own wrongdoing. I can't help but think as well of the barred owl's blood on the snow and the relief of the timber companies and the forest managers to have found a way to continue to cut the big trees, to continue to destroy spotted owl habitat, to continue to create a patchy landscape of edges and blowdown where barred owls thrive—and blame the owls. Cripes. It is part of a lethal response that echoes all over the American landscape—to frantically fight back against the *effects* of destructive human decisions rather than to address the *causes,* the destructive decisions themselves.

RESILIENCE

Resilience is defined as "the ability of something to return to its original shape after it has been pulled, stretched, pressed, bent, et cetera." I've seen PowerPoint presentations of a hammer bouncing onto a rubber sheet; the hammer flies off and the sheet vibrates back into position. Or this: A ball rolls down a J, swings up the curve, balances on the tip of the little tail as if it might drop off into alphabetical oblivion, and then rolls back down into the embrace of the curve. The point is an important one. Ecosystems, social systems, and moral systems have been "pulled, stretched, pressed, bent, et cetera" by destructive patterns of living. How much distortion can they take until they snap, no longer returning to something resembling themselves? And given that there is a limit to the distortion any system can take, how can its resilience be increased?

I can understand the wide concern for resilience: Don't we all wish in our hearts that things could snap back into some semblance of the way they used to be? We wish for resilience in our marriages, our knees, our teenagers; and we wish for resilience in heron rookeries, coral reefs, salmon streams, that they might take terrible hammer blows and bounce back again. And if we could engineer this resilience into our economic or ecological systems, what a load that would take off our backs and consciences. Resilience: the power to rebound (*re-*, "back," plus *salire*, "to jump, leap").

Maybe in some cases, this is possible. But in the new world's increasingly acidic oceans, chaotic climate, and disintegrating life-supporting systems, there may be no going back. If that is even partly true, the challenge of the coming time is what I will call *presilience*, literally, *forward-jumping*: the courage to take a great, stumbling leap into a world unlike any we have ever seen, knowing that we will not be returning to the old ways.

SOME YEARS AGO, I visited Baranof Island in Southeast Alaska for a conference on community resilience. The people were understandably concerned. There are two ways to get to the island—by air or by boat. Food arrives on the island in one way—by boat.

Fuel arrives on the island in one way—by boat. Wealth arrives on the island by boat—fishing boats, or glittering cruise ships holding four thousand tourists. Not very resilient, this business of putting all your eggs in one basket. A breakdown in transportation, and it's not many weeks before the fishing boats are out of diesel and the grocery shelves are empty. But it's not just this island whose resilience is brittle. Anything that strips down a system, simplifies it, homogenizes it, mass-produces it, increases its vulnerability and decreases its ability to endure through change.

But putting all our eggs in one basket has been the world's project for the last few centuries, has it not? The industrial growth economy has narrowed and narrowed future options by building infrastructure for the exclusive use of fossil fuels while undermining competitive sources of energy. It has dramatically reduced biodiversity among living things. It has eliminated cultures, languages, indigenous lifeways, and lives, and replaced them with the global economy. It has grown one genetically modified variety of corn and lopped the heads off any stalk of wheat that grew to a nonaverage height. It has made sure that each Big Mac is exactly like the other 47 billion, even as it demeans any ways of loving or living that differ from the "norm" and measures all value in U.S. dollars. It has tried to put all the world's eggs in one basket, and now it sings, over and over again, the same dumb song ("I'd Like to Buy the World a Coke").

Let's extend the metaphor: What did the farmer do after he put all his eggs in one basket and then tripped over a hay rake?

Here's what I see, a scene rolling in slow motion: The farmer hurries across the farmyard, whistling. A basket of eggs swings from his hand. Because his eyes are on the kitchen door, he doesn't see the hay rake. He trips and sails forward—a giant leap. His legs pedal air. The hand holding the basket reaches forward, as if it were holding a lantern. The eggs bound out of the basket, one after another—twelve eggs flying. And one after another, they flatten against the pounded earth. Soon after, the farmer's body smacks full length onto the ground. There is stunned silence in the barnyard. Then, wide-eyed, the man lifts his head. He has egg on his face, and blood in his nostrils.

No point in belaboring this; we all breathe through the smell of blood. More important, what can the analogy teach us? Mark Twain famously wrote, "Put all your eggs in the one basket and—WATCH THAT BASKET." This is the strategy pursued with lethal seriousness by those in power. The more perilous the single path, the more viciously they insist on it. Consider the fossil fuel industry's stranglehold on Congress. Consider the attacks of the solipsistic self-righteous on working women, gays, black voters, the desperately poor, and immigrants (take your pick; this is a war against difference, which takes many forms). Consider the relentless campaigns against climate science, against all science. Consider the concentration of well-protected money. Consider the hegemony of the individualist, capitalist worldview.

But listen, everybody: Mark Twain was a very wise man, but "Put all your eggs in the one basket and watch that basket"? That was a *joke,* for god's sake. He wasn't *serious.*

There's a different lesson to learn, an obvious one: We need lots of baskets, a wild abundance of baskets. Willow baskets, pink straw baskets, grocery baskets, baskets woven of spruce roots in the dark winter or carved from stones and stories. We need to cherish and protect them all. And eggs? How many kinds of eggs can we find or create? Lizard eggs, spider eggs carried on the wind, golden eggs laid by geese, all the embryos in dark places—seeds, always seeds, and especially the seeds of new ideas. Bicycle co-ops, farmers' markets, collaborative villages, neighborhood windmills, shared jobs, repair shops, local money, recyclable houses, expanded families, wildlife refuges in the parking lots, gardens in the front yards, book exchanges on the corner, songs, always songs. If it holds life, nurture and protect it in every possible way. That will at least give humanity a chance to answer the world's desperate call for new ways of living together with the good Earth.

This is the work I saw under way on Baranof Island. Some of the islanders were practicing presilience, even though they did not call it that. In meetings and workshops, they were learning how to raise vegetables in the cold and pull power from the rain and, most important, bring their minds together to radically imagine lasting ways of life. In this, they have the inestimably valuable legacy of a thriving Tlingit community, which has lived on the island called Sheet'-ká X'áat'l for more than ten thousand years. Other islanders, however, are still steadfastly guarding the basket, working for more of the same—more salmon, more tourists, more barges—even though not many of these will be floating on the warmer, sourer, stormier seas that scientists predict when they look into the new world.

I wandered through that week on Baranof Island in a rain of helpless weeping. This doesn't often happen to me, and when it does, I don't usually talk about it; I take pride in my spine. But the irretrievable losses overwhelmed me that week—and the wobbly assumptions that the salmon, the tourist industry, and the transportation system can bounce instead of shatter. Shatter: a rain of fish scales and FOR SALE signs. I walked and walked through dark rain while ravens *clonk*ed and Orthodox crosses tipped deeper into beds of moss. All I could do that week was walk in wet darkness, half-blinded by rain.

It was months before I was able to think about this important fact: that a forward-looking community, wildly imagining together, can help determine what the basket holds.

Here's what I wish for the world's baskets:

1. The greatest possible abundance of living things, who hold in their DNA, in their wings and eyestalks (and in the tangled connections among all beings), an infinite and never-ending variety of ideas about how to thrive in changing conditions.

2. The greatest possible diversity of human beings, who hold in their stories—their lifeways, their hard-won wisdom, their languages, their lived songs of love and grief—an irreplaceable heritage of ideas about mutual flourishing, even at the ends of the Earth.

3. The greatest possible abundance of fresh, clean water in a million ponds and rivers, immortal ice, dependable rain.

4. Layers of fertile soil—the beautiful material condition of abundant life.

5. The greatest variety of tools and skills of every kind (how to catch a codfish or calm a child, how to fix pumps or can beets).

And, most of all,

6. Respect for all these, abiding love for all these, and the moral courage to nourish them without rest or fear, knowing that our baskets hold the origins of the next lives.

DESPAIR

And so we come to the last item in my list of false promises and dead ends. In a few more chapters, we will get to the subject of the work to be done. But right now, the subject is despair, which is the dead end to end all dead ends. Etymology would have us believe that it's a nothing, an absence, *de-* (without) and *sperare* (hope). But I believe despair is very much a thing. It is an absence that has become dangerously alive—a dark, damp, descending thing, capable of enveloping a person and stopping all action. Not an emotion; by definition, emotions excite movement. It's the opposite of emotion: the leaden feeling of nothing left to feel. Despair is, thus, losing heart.

When we think about the work of stopping the environmental collapse, of course we feel despair, you and I. Our options are limited. Cities and homes and transportation systems are disgracefully designed, destructive ways of living are skillfully protected by tangles of profit and power around the world, corporations are behaving like psychopaths, congressional leaders are behaving like children, public attention is distracted by war and football, and we have run out of time. The consequence of failure is the end of civilization, perhaps the miserable, moaning end of human life. How's that for reason for despair?

The lyrics to Otis Redding's 1967 hit come into my mind, complete with its seagull whistles and rocking waves:

> So I'm just gon' sit on the dock of the bay
> Watchin' the tide roll away . . .
>
> Looks like nothing's gonna change
> Everything still remains the same
> I can't do what ten people tell me to do
> So I guess I'll remain the same . . .

But, at the exact moment that despair is intellectually justified, it is morally impossible. The very facts that paralyze action are the facts that require it: We love the lives on this planet, and they are being taken from us.

This is why thought leaders of all sorts insist, in every way they can, that we must—as an act of will and courage, as an embodiment of our humanity—turn away from despair. The words of author Clarissa Pinkola Estés ring true to me:

> Do not lose heart. We were made for these times. . . .
> Yes. For years, we have been learning, practicing, been in training for and just waiting to meet on this exact plain of engagement. . . .
>
> I recognize a seaworthy vessel when I see one. . . . There have never been more able vessels in the waters than there are right now across the world. And they are fully provisioned and able to signal one another as never before in the history of humankind.

> . . . There will always be times when you feel discouraged. I too have felt despair many times in my life, but I do not keep a chair for it. I will not entertain it. It is not allowed to eat from my plate.
>
> . . . I hope you will write this on your wall: When a great ship is in harbor and moored, it is safe, there can be no doubt. But that is not what great ships are built for.

What strikes me about her admonition is the assumption that one can turn despair away, speak sternly to it as if it were a begging dog, and send it from the table. I think this is true, but I also know it is not easy. How does one find heart? That's the question: How does one find heart?

I have three thoughts about this.

One answer is paradoxical: Grief is a powerful antidote to despair. Grief is an emotion, a terrible aching emotion, but it is something rather than a powerful absence. *Grief,* from *gravis,* a heavy weight, a burden almost too great to bear. A person in despair is well advised to open her heart to grief, the weight of sorrow. Let it pump like lifeblood through the broken heart. A bear swimming forever northward to drown, a thin old man dreaming about corn pudding, a fledgling wren with its wings on fire, the small voices of bewildered children murmuring themselves to sleep in a church basement after the storm. It's so sad. "It's just sad, is what it is," my evangelist friend said, and we cried together on the front porch.

The human response to *despair,* it seems to me, is drugs or death.

But the human response to *grief* is creation, often the creation of beauty. "We have art," philosopher Friedrich Nietzsche wrote, "in order not to die of the truth." After a friend dies, there is a compassionate letter written, a casserole made, a coffin crafted from cherry wood, a tree planted. There is music—the heart-opening chords of old hymns, Rachmaninoff's hollow descending bells, Odetta's blues, dancing people in the dusty field, the passions of Bach. There are pietàs carved from marble or tombs engraved on desert sand or roadside shrines of plastic flowers, teddy bears, and ribbons. Often, humans respond to grief with the gritty beauty of collective action—the mourning crowd that gathers at the church after the funeral of four little girls killed by a bomb.

And humans sometimes respond to grief by turning to the comfort and reassurance of the natural world, its peace: the steady surge and flow of the sea on sand, water slipping over stones. Sorrow is part of the Earth's great cycles, the surge from living to dying to life again. Maybe this is why grief can make a deeper connection to the currents of life and so connect, somehow, to sources of solace and courage. Maybe, in these times, it will be grief that saves us from despair.

Maybe. But work also is a powerful antidote to despair. That's my second thought. "If you're in despair," environmentalist Bill McKibben said, "you're not working hard enough." I think that's right, and that has been my experience. My sister, who is a kindergarten teacher, says that she will never tell children about the harm befalling the world without telling them what they might do to help. The drawings of the rain forest animals that they send to the newspaper might not save the world, but they might save the children from falling into inaction and despair. In my own university classrooms, I have learned that the more students learn about climate change, the less despair they feel. How could this possibly be? Because they have taken the action to learn what they can, and by that action, they have reclaimed their power to make change and have, not incidentally, found themselves in a community of empowered and caring people.

I'm having trouble with my third thought about how to turn away despair. It's an answer that is urged on me by my dearest friends. They urge me to be present to the beauty and joy that is in the world right now, this timeless moment. That doesn't require me to set aside thoughts about the horror of its peril. In fact, "being present to beauty and joy," the poet Charles Goodrich wrote to me, "deepens the capacity for grief, anger, confusion—and maybe clarity, too, even if it's clarity about unthinkable futures and irresolvable dilemmas." But the obligation to be grateful for the gifts of the Earth calls me to rejoice in the beauty that is present to me right now, calls me to *en-joy* it.

The subsistence poet Hank Lentfer agrees with Charles. Opening one's grieving heart to the beauty and joy of this world, right now, is an answer to despair and a moral obligation based on love. Do you love a person less who is dying? he asks. On the contrary, every moment with that person takes on more value, a deeper joy, a baffling beauty—because of, not in spite of, the fact that she will soon pass from this world.

Yes. Yes. This makes sense to me, and god knows I am desperately hungry for that direct apperception of the sweetness of the moment. I live for that. But here is the weird truth: I feel uncomfortable being glad for the beauty and joy of the moment, especially in those cases when I am fully aware that the beauty is the direct result of ongoing and pending destruction. Here is a story about my struggle:

NEW BEAUTY IN THE RUSHING CHANGES

It's been a beautiful summer on our Alaskan inlet. No one can remember a summer so clear and warm for so many days. The children headed down to wade on the gravel beach yesterday afternoon, but before long, they were on their stomachs in the shallows, propped on their arms, kicking their feet and gurgling like floatplanes. How they so quickly learned to do that, I don't know. This is the first time in their short lives that the water in Alaska has been warm enough to tempt the children to swim. Parents sat in the sun on the beach with a carrot cake and a jug of lemonade between them, never worrying about bears. The bears just haven't been around this summer, even though pink salmon are circling at the mouths of the streams. Fishermen on the seiners are ecstatic. They can encircle and net a ton of fish with one haul. Even better for the fishermen, there are no whales in the inlet, tangling the nets the way they did last year. The water itself is beautiful, clearer than anyone remembers, an azure blue some of us recognize from the Caribbean, but never before from the inlet. The air is sweet with spruce sap.

Such a treat, this string of warm days and clear nights, one after another after another after another all summer, and every person on the trail says, "Such a beautiful day. What did we do to deserve this?"

And then they keep walking, because they have some answers to that question, or at least some reasonable suspicions. The freshwater flowing into the inlet is warmer because the ancient spruces whose green shade once cooled the streams have been cut for lumber. The seawater is warmer because it's capturing the heat of the supercharged atmosphere. The salmon are circling in front of the streams because the streams don't have enough water to rise over the gravel bars that block their entrance. There's not enough water because the snowfields melted and drained into the inlet long before the salmon arrived, and it hasn't rained since. The clear water? My neighbor thinks it's the low stream flows that no longer flush nutrients into the coves. No one is really sure what happened to the bears, or why they were roaming the forests in January, when they should have been hibernating. Whales are missing because whales follow baitfish, and there haven't been many herring balls in the inlet this year. The warm weather stalled here because the great global wave of the jet stream, which once reliably brought storm after storm, has been wrenched from its tracks; it is taking our storms to other lands. The sweetness on the air is the smell of dry forests dying.

I sit on the beach and watch the happiness of the children, brought to them in part by the global warming that, unless we find a way to stop it, will irretrievably damage their lives and prospects. I hear the seiners, their winches grinding, and watch the ballet of mother ship and tender as they net the last of the salmon run. Flashing bodies pour onto the steel decks. Every change, no matter how pleasant, has become a poke to the heart. Is this the beginning of the end? Is this the crest of a wave poised to plunge into a darker world? I don't know what to think.

Into the center of my confusion, Hank sent me an email message: "I hope you're finding some hints of new beauty amid the rushing changes," he wrote. "Henceforth, I guess, that's one of our tasks. Whether our species can stay around to extol them remains to be seen, of course."

Yes, I will tell him honestly. There are more than just hints of "new beauty amid the rushing changes." The rushing changes are, without a doubt, bringing new beauty, full-blown. I will tell him that on a distant island, just last week, I found the entire skeleton of a bald eagle. He lay on his back with his wings spread across a bed of moss. The bones were picked white. But the flaring flight feathers were intact, the white tail feathers undisturbed. In his skull, the eye sockets and heavy beak pointed east. His knucklebones were like amber stones. I don't know what caused this lonely death, what storm, what hunger. All the same, there in that shadowed and mossy hollow, with light from the ocean glancing across the green sweep of the hemlocks, the skeleton was beautiful in ways that were new to me.

When, clambering up behind, the children caught sight of the skeleton, they stopped and let out their breath. *Whoa*, they said in little flute voices. Sea light wavered on their salty hair and flashed across their life jackets. What I saw then were two small children standing silent and uncertain in the presence of a death of great import. Maybe this scene is an example of the new beauty.

But how can it be one of my joys, to celebrate the new beauty to be found in the rushing changes, when I know that those rushing changes are ruinous, and when I know that the same changes that have brought us a beautiful summer in southeast Alaska have brought violence to the sweltering cities, wildfires to the west, and suicide to arctic coastal towns? Beauty is good. And this beauty is awful. With starshine and feathers, it masks a hideous face. Whether the "new beauties" are beautiful is beyond question. Dear god, they are glorious. My question is whether accepting the new beauty as a gift is shameful in some way, an accommodation?

The agony is that each of us is living in two worlds at once, and each world calls for a different set of emotional and moral responses. The first world is the one I wade in, holding small children by the hand, stopping to admire every seashell and name the color of its stripes. The second world is a shadow of the first, made of the absence of whales, the spars of mysteriously dead spruce, limestone mountaintops bare of snow, a new disease that seems to be draining the starfish of whatever holds them rigid, rotting salmon undisturbed by bears, fear for the children, Wendell Berry's "forethought of grief." The first world is made of love and astonishment. The second is made of knowledge. The first calls for joy and wonder, the second for outrage and resistance. I'm tempted to say that the first is the world of the child, the second, the world of an adult.

Most days, I can walk well enough in the world of laughing children. But I keep slipping through the thinning ice of it, losing my footing and plunging into despair at what the world will soon be. I dangle there, because I have no tools, no intellectual ice ax, no emotional grappling hook. How does a person live in two worlds at once? I'm looking for some example, some wisdom from human experience.

Maybe this: In the time of mysteriously silent springs, Rachel Carson warned that we lived in two worlds. One was a world of miracle pesticides, perfect apples, gardens spilling with petunias, and picnics by the river, an idyllic time after the war when children could play outside in the evening and never worry about mosquitoes. The other world was hidden in darkness, when trucks rumbled through the suburbs, spraying poison fog into the streetlights, when robins fell on their backs, convulsing. I suppose it was morally possible to live in that perfect world, but only for a time, before the full sorrows of the hidden world revealed themselves and people understood the true cost of the laughter of children playing hide-and-seek in sweet dusk. Then, the danger was that the enjoyment of that beauty might allow a person to discount the dying birds as a price to pay for such pleasure, and so avoid responsibility for stopping the agents of silence and death.

I'm afraid that this might happen as global warming accelerates: that, over time, more and more of us will come to live unknowing and maybe contented in the world of what remains, and will not know the world in its agony of rushing change. We then could be happy amid new beauties, unaware of the limited perception that hides our own tragedies and those of the world.

But I could have this all wrong, because there's another, a simpler, example of living in two worlds that every person who's lived on the water knows. I've stood often enough with one foot on the dock and another on the bow of the skiff to know the perils of standing with a foot in each of two worlds. The lesson here is that you have to commit. The boat is leaving, and you are going with it gracefully, or you are going under. Unless, unless you can hold on to both boat and shore. This takes balance, and it takes strength in the knees and inner thighs to hold the boat with one leg against the wind that pries the skiff from the dock.

Hank and Charles are telling me that the boat is briskly sailing away from the world we knew, leaving our snug cabins and gardens behind, the zucchinis, rhubarb, and stacked firewood abandoned and pathetic. Look to the rainbow in the spray off the bow, they are saying. Look to the platinum reflection of clouds on the swell. Look to the lavender stripes in the sunset. Listen to the song in the sails. Isn't this a measure of your love for the world—that you will find it beautiful, even when it is desperately wounded? Find the new beauty in the rushing changes. In whatever direction you are heading, find beauty there. Find happiness. This is the solace open to you: that no matter what happens next, Earth is spectacularly beautiful, and it will be spectacularly beautiful whether we humans sail off the edge of the planet or not.

I know they are partly right. This is a beautiful and beloved world, and I will accept the gift gratefully and with lasting love. If this world is full of holes and heartbreak, then let that be my world. I will live in it fully and love what it offers. At the same time, I will not forget what the world has lost and stands to lose. The precarious balance between joy and sorrow will be my answer to despair. I will hold the exquisite filigree of a dead starfish in one hand and grip the strap of a child's life jacket with the other. Together, the child and I will watch the moon rise above the last of the alpenglow on the mountains.

the work of democracy

PLATO WAS NOT a fan of democracy. Most people are not attentive to the art of government and are not particularly interested in educating themselves, he wrote in *The Republic.* Some withdraw from the affairs of state entirely, not voting, not caring, just attending to their own small interests; these are the *idiotes.* Most, ruled by their emotions and so easily deceived, can be lured into supporting stupid wars or larcenous politicians.

The oligarchs and plutocrats, the few and the wealthy, take good advantage of the citizens' weaknesses to seize power. It's easy enough to fool most people with appeals to fear or hatred; it's easy enough to buy their votes or simply ignore the law and take what you want. So democracies always turn into oligarchies, Plato believed. Oligarchies, for their part, inevitably become anarchies. That's because the people—unjustly and brutally impoverished, slaughtered in wars that profit only the wealthy—rise up in despairing anger and destroy the government.

This was Greece around 380 BCE. But damn, it sounds familiar. The United States has clearly made the transition from a democracy to a plutocratic oligarchy, and for all the reasons Plato cited. The big question is not whether the oligarchy will last. It will not;

injustice is inherently unstable. The big question is whether the oligarchy will be dismantled by an orderly return to democracy or by rapid and irreversible anarchy. I think it has come to that. Either way, the people will not tolerate forever a system that—let's lay it out—(a) forces them to bear the risks of the recklessness of a few powerful profiteers, to assume the burdens of others' privilege, and to pay the real costs of destructive industries in the currency of their health and the hopes of their children and (b) in the process, threatens to disrupt forever the great planetary cycles that support all the lives on Earth.

Climate change. Toxic neighborhoods. Financial recklessness. Jobs despair. Concentrated wealth. Wholesale extinctions. Pointless war. It escapes no one's attention that all these connect to a central social pathology, which is the funding (one might say, the buying and selling) of elections (and of the elected) by powerful centers of wealth: mostly corporations, mostly destructive and extractive corporations. Our erstwhile democracy has now developed a futures market in politicians. This has created a situation in which the government is fundamentally controlled by those who would risk or wreck the (name your favorite: economy, environment, children's futures) for their own short-term gain.

The consequence is, of course, that the destructive few are moving into position to control the regulatory agencies and potential regulations that might have limited their recklessness and greed. They have the consequent power to close off options for resolving the environmental and economic emergencies. They have the power to block federal actions that might prevent injustices. They have the power to bulldoze the natural systems that sustain our lives. *We need smaller government*, the oligarchs cry, which is code for, *We need a government that is toothless enough to allow unregulated corporations to raid and degrade the commons without consequence.*

Self-created environmental catastrophe has taken down many civilizations before ours. But this time, the self-inflicted catastrophe of climate change will take down also the hydrological cycles and relative climate stability that have allowed the evolution of the world as we know and love it. The time has passed for an environmental movement. The time has passed for a climate change movement. The time has passed for isolated grassroots movements. We stand on ground that trembles with tectonic movement. Along the straining fault lines of our civilization, we feel the building forces for justice, sanity, and lasting ecological and cultural thriving. When it comes, this will be the Big One—the coming together of all of us who care about the future and do not want to gamble it away. The Big One will shake the world.

I CAN'T SAY I'm a fan of Plato. He began *The Republic* by claiming that only three types of people are born in the world: the bronze, the silver, and the gold. The bronze people do the work of slaves. The silver people do the work of warriors. And the golden people do the work of ruling. In fact, in Plato's ideal republic, the rulers are philosophers. This, I believe, is a truly terrible idea.

No. When I want to get a good dose of practical wisdom about the work of democracy, I turn to a desert rat and canyon writer, Edward Abbey. He's the author of *The Monkey Wrench Gang, Desert Solitaire, Down the River,* and many more books very much worth reading. How desperately the West, and the wild, and the world need Edward Abbey today.

What would he say about Congress? What would he make of the stamper trucks shaking red rocks off the cliffs? What fine adjectives would he find for the drilling rigs on fracking pads spattered all the way from the Uinta Basin to the San Juan River? What would he say to the oil company executives who are cheerfully taking down the great natural systems that sustain life on the planet to jack up profits that are already the highest since the pharaohs? How would he grieve for his beloved spotted toads, besieged by global warming, pipelines, and roads? What would he say when he climbed to the hidden springs where for centuries the mountain lions have lain down with the lizards, only to find the water gone, poisoned and forced underground to fracture the rock to release the natural gas to earn a fat CEO $22 million a year? What would he say about the silent masses, the corrupted science, the solipsistic consumers and sociopathic corporations? What could he tell us gloomy citizens about our moral responsibilities in a time of global warming, ecosystem collapse, and wholesale government failure?

But Ed's gone, either gently desiccating in a desert grave or swearing softly in heaven, having ascended with his soul tied tightly to the soul of Leslie McKee's wife, a kind Mormon woman who promised she would carry him heavenward when she herself went. All the same, his fierce love of the world is still with us in his rampaging books and letters. I found him in Powell's Books in Portland, Oregon, dog-eared and leaning slightly to the left, in a crowded section labeled NATURE. Then, with some serious slicing and splicing, I deciphered his answers to the questions that have been haunting me.

Herewith,

A POSTHUMOUS INTERVIEW WITH EDWARD ABBEY

KDM: If you drove a nail through the center of this decade, all of planetary history would swing in the balance. Global warming, massive extinction, acidification of the oceans, poisonous air and land and water—all these have brought us to a hinge point in history. In this decade, we will lose it all, or we will redeem a just and thriving planet. Just telling you. Do you have words to describe this peril?

EA: *Humanity has entrapped itself in the burning splendor of* technikos.[1] *We are cursed with a plague of diggers, drillers, borers, grubbers, of asphalt-spreaders, dambuilders, overgrazers, clearcutters, and strip-miners whose object seems to be to make our mountains match our men—making molehills out of mountains for a race of rodents—for the rat race.*[2] *My god, I'm thinking, what incredible* shit *we put up with most of our lives.*[3] *The present course is one of premeditated suicide.*[4] *We are befouling and destroying our own home, we are committing a slow but accelerating life murder—planetary biocide.*[5]

KDM: You're a well-credentialed philosopher, Mr. Edward Abbey, MA from UNM. Philosophers believe in the power of ideas to shape history. We know that it's a crackpot cosmology that got the world into this scrape: a worldview that brags that humans are separate from and superior to the Earth, in charge and in control—and that the planet and all its lives have no value except their usefulness to

---

1 Edward Abbey, *Down the River*, NY: E.P. Dutton, 1982, p. 93.
2 *Down the River*, p. 34.
3 Edward Abbey, *Desert Solitaire*, NY: Touchstone, 1990, p. 155.
4 *Down the River*, p. 105.
5 *One Life at a Time, Please*, p. 177.

ours. We need a new set of answers to the foundational questions of the human condition.

Do you know the answers?

EA: *Though a sucker for philosophy all of my life I am not a thinker but—a toucher. I believe in nothing that I cannot touch, kiss, embrace—whether a woman, a child, a rock, a tree, a bear, a shaggy dog. The rest is hearsay.*[6] *What else is there? What else do we need?*[7]

Religions, all of them, tend to divorce men and women from the earth, from other forms of life.[8] *We are obliged to spread the news, painful and bitter though it may be for some to hear, that all living things on the earth are kindred.*[9] *We are kindred, all of us, killer and victim, predator and prey, me and the sly coyote, the soaring buzzard, the elegant gopher snake, the trembling cottontail, the foul worms that feed on our entrails, all of them, all of us.*[10]

KDM: But if all of us are kin, then surely we don't have the right to treat every creature in the whole buzzing world like chattel slaves, as if they belonged to us to ruin or to auction: coyotes and gut-shot cactus, entire fields of mice and crows. What would you call for instead?

EA: *Recognition of the rights of other living things to a place of their own, a role of their own, an evolution of their own not influenced*

6 *Down the River*, p. 57.
7 *Desert Solitaire*, xiii.
8 Edward Abbey and David Peterson, *Postcards from Ed*, Minneapolis, MN: Milkweed Editions, p.93.
9 *Desert Solitaire*, p. 21.
10 *Desert Solitaire*, p. 34.

by human pressures. A recognition, even, of the right of nonliving things—boulders, for example, or an entire mountain—to be left in peace.[11] *In demanding that humans behave with justice, tolerance, reason, love toward other forms of life, we are doing no more than demanding that humans be true to the best aspects of human nature.*[12]

KDM: The kinship metaphor suggests that we and our brothers and sisters on the land share a common home. What does that tell us about what we ought to do?

EA: *With bulldozer, earth mover, chainsaw, and dynamite, the international industries are bashing their way into our forests, mountains, and rangelands and looting them for everything they can get away with. This for the sake of short-term profits in the private sector and multimillion dollar annual salaries for the three-piece-suited gangsters.*[13]

If our true home is threatened with invasion, pillage, and destruction—as it certainly is—then we have the right to defend that home, by whatever means are necessary. We have the right to resist, and we have the obligation. Not to defend that which we love would be dishonorable.[14]

Be of good cheer, the military-industrial state will soon collapse. Meanwhile, we must do all in our power to oppose, resist, and subvert its desperate aggrandizements. As a matter of course. As a matter of honor.[15]

---

11 *Down the River*, p. 119.
12 *Postcards from Ed*, p. 94.
13 *One Life at a Time, Please*, p. 30.
14 *One Life at a Time, Please*, p. 31.
15 *Down the River*, p. 4.

KDM: Really, by whatever means are necessary? Killing?

EA: *Not people. We're talking about bulldozers. Power shovels.*[16]

KDM: Breaking the law?

EA: *Why not?*[17] *You think this is a picnic or something?*[18] *The "choice-of-evils" statute allows the intentional commission of an illegal act when the purpose of such act is to prevent a greater harm or a greater crime.*[19] *"Protest is always justified," said [physicist and nuclear activist John] Gofman, "when it is the only means to make a deaf government listen."*[20]

KDM: But blowing up a railroad bridge, pushing a loader into the drink, as in *The Monkey Wrench Gang*? These days, people would call that ecoterrorism, and it'd be a long time before any one of your heroes passed a bottle around the campfire again. Police don't like gang members either; they shoot them.

EA: *Let's have some precision in language here: terrorism means deadly violence—for a political and/or economic purpose—carried out against people and other living things, and it is usually conducted by governments against their own citizens or by corporate entities against the land and all creatures that depend upon the land for life and livelihood. A bulldozer ripping up a hillside to*

---

16 Edward Abbey, *The Monkey Wrench Gang*, NY: Avon Books, 1992, p. 58.
17 *The Monkey Wrench Gang*, p. 101.
18 *The Monkey Wrench Gang*, p. 77.
19 *Down the River*, p. 103.
20 *Down the River*, p. 109.

strip mine for coal is committing terrorism. Sabotage, on the other hand, means the use of force against inanimate property. The characters in Monkey Wrench *do this only when it appears that all other means of defense of land and life have failed and that force—the final resort—becomes morally justified. Not only justified but a moral obligation.*[21]

KDM: We've got to talk about moral obligation. You wouldn't believe how many people treat the environmental emergencies as just technological or scientific problems, economic problems or even national security problems. But honestly, taking whatever you want for your profligate life and leaving a dangerous and ransacked world for the next generation is a moral failure, and it calls for a moral response. You're a philosopher: What are some of the moral reasons why we have to push back against the forces of destruction?

EA: *What I am concerned about is the world my children will have to live in, and maybe, if my children get around to it, the world of my grandchildren.*[22]

KDM: That is a good consequentialist argument for action, based on the hope that it might make a difference to the people you care about. But you write so often about honor and dishonor that I wonder if you're working toward a virtue ethic—judging the rightness of actions not by whether they save the world (they probably won't) but whether they spring from a virtuous character . . .

---

21 *Postcards from Ed*, p. 129.
22 *Postcards from Ed*, p. 93.

EA (interrupting): *and high moral purpose—concern for right and wrong, justice and injustice, truth and falsehood, beauty and ugliness.*[23]

KDM: And an obligation to be loyal to the Earth?

EA: *Loyalty to the earth, the earth which bore us and sustains us, the only home we shall ever know, the only paradise we ever need—if only we had the eyes to see. Original sin, the true original sin, is the blind destruction for the sake of greed of this natural paradise which lies all around us—if only we were worthy of it.*[24]

In any case, the beauty and existence of the natural world should be sufficient justification in itself for saving it all.[25]

KDM: Now you're talking as though the natural world has intrinsic value, value in and of itself, not just because it's useful to us. Are you affirming that the natural world should exist for its own sake, for its beauty and mystery and wonder?

EA: *You bet, Doc.*[26] *Its significance lies in the power of the odd and unexpected to startle the senses and surprise the mind out of their ruts of habit; to compel us into a reawakened awareness of the wonderful—that which is full of wonder. For a few moments, we discover that nothing can be taken for granted, for if this ring of stone is marvelous then all which shaped it is marvelous, and our journey here on earth, able to see and touch and hear in the midst*

23 *Down the River*, p. 9.
24 *Desert Solitaire*, p. 167.
25 *Down the River*, p. 120.
26 *The Monkey Wrench Gang*, p. 179.

of tangible and mysterious things-in-themselves, is the most strange and daring of all adventures.[27]

KDM: What is the special obligation of a writer in this astonishing but desperately wounded world?

EA: *I believe that words count, that writing matters, that poems, essays, and novels—in the long run—make a difference.*[28] *The writer's job is to write, and write the truth*[29]*—especially unpopular truth. Especially truth that offends the powerful, the rich, the well-established.*[30] *But he also has the moral obligation to get down in the dust and the sweat and lend not only his name but his voice and body to the tiresome contest. How far can you go in objectivity, in temporizing, in fence-straddling, before it becomes plain moral cowardice?*[31] *"It is always the writer's duty," Samuel Johnson said, "to make the world better."*[32]

KDM: But what is your vision of a better world? It's easier to imagine the end of the world than the end of business-as-usual. The financial power of the fossil fuel industry has really done a job on our democracy. But maybe that doesn't matter; Plato didn't think much of democracies, anyway. Democracies always turn into plutocracies, the rule of the rich, he said, because rich people can buy votes. And plutocracies always turn into anarchies, because the

---

27 *Desert Solitaire*, pp. 36-37.
28 *One Life at a Time, Please*, p. 162.
29 *Postcards from Ed*, p. 214.
30 *One Life at a Time, Please*, p. 163.
31 *Postcards from Ed*, p. 215.
32 *One Life at a Time, Please*, p. 178.

people won't stand for injustice forever. So that's the vision we're left with? Anarchy?

EA: *Anarchy is democracy taken seriously. An anarchist community would consist of a voluntary association of free and independent families, self-reliant and self-supporting but bound by kinship ties and a tradition of mutual aid.*[33]

KDM: I just don't know how to get there. It's hard for activists to find a focus these days. Already the world is knocked off kilter by big storms, rising water, wars in desert lands, starving people on the move. What kept you going? What powered your steadfast efforts to defend the land?

EA: *Equal parts anger and love.*[34]

KDM: But how did you keep from falling into utter despair?

EA: *Thoreau said, "Who hears the rippling of rivers will not utterly despair of anything." That makes sense.*[35] *Where there is no joy, there can be no courage, and without courage all other virtues are useless.*[36] *One single act of defiance against power, against the State that seems omnipotent but is not, transforms and transfigures the human personality.*[37] *The search for transcendence and*

33 *One Life at a Time, Please*, p. 26.
34 *One Life at a Time, Please*, 176.
35 *Down the River*, p. 3.
36 *Desert Solitaire*, p. 125.
37 *Down the River*, p. 108.

integrity and truth goes on.[38] *Best to march forth boldly, with or without life jackets, keep your matches dry and pray for the best.*[39]

KDM: I don't know where you are now, Edward Abbey. But I hope that maybe some of us citizens (of the United States, of the planet) might find you in ourselves—your bravura, your common sense, the ferocious truth, the anger and the love that will empower us to save your "unpleasant solpugids" and your slickrock—and maybe even save our own sorry skins.

All the mining and drilling, the burning, burning, the thunder and the roar, all the oily gases, and now the melting planet with its crazy storms, the toxic water and multiplying cancers, all the dying, dear god, the unnamed creatures, the singing frogs, the children of the future, whose small voices cannot call out to us. It's moral monstrosity. We can't allow it. From your place in the stony cycles of the planet, what is your advice to us?

EA: *It's time to get f*cking back to work.*[40]

---

38 *One Life at a Time, Please*, p. 216.
39 *Desert Solitaire*, p. 240.
40 *The Monkey Wrench Gang*, p. 123.

the work of science

JAMES HANSEN, ONE of the world's foremost climate scientists, retired from his position at NASA's Goddard Institute for Space Studies after a forty-six-year career so that he could take a central role in climate activism. Already arrested four times in demonstrations, Hansen is now determined to sue the U.S. government over its failure to take stronger action to slow climate change. "Dr. Hansen's activism of recent years dismayed some of his scientific colleagues, who felt that it backfired by allowing climate skeptics to question his objectivity," the *New York Times* reported. "But others expressed admiration for his willingness to risk his career for his convictions."

DAVID SUZUKI, PHD, zoologist, professor emeritus at the University of British Columbia, stood on the steps of the Toronto Courthouse. "Governments and corporations are not just failing us," he declared. "They are the driving forces that are taking us to the brink, willfully ignoring the consequences and thereby committing what can only be called an intergenerational crime. Willful blindness is an indictable offense . . . our so-called leaders must be held accountable."

Then he called on Canadians to try him for seditious libel—a crime that he could commit only if he were lying. Let the Canadians decide if he was telling the truth or not, he said.

Complex systems researcher Brad Werner gave a talk at the American Geophysical Union this year called "Is Earth F*cked? Dynamical Futility of Global Environmental Management and Possibilities for Sustainability via Direct Action Activism." Global capitalism has made the depletion of resources so rapid, convenient, and barrier-free that Earth-human systems are becoming dangerously unstable in response, Werner said. He called for "environmental direct action, resistance taken from outside the dominant culture, as in protests, blockades, and sabotage by indigenous people, workers, and other activist groups." It is not clear whether he was including scientists in that call.

"When you are wearing your scientist's hat," the vice president for research said, "then you should be communicating science. This does not mean that you can't take off your scientist's hat, put on your citizen's hat, and advocate particular policies. Just be sure to separate your policy views from the positions of the university and the conclusions of science."

In 1937, the Prussian Academy of Sciences accused Albert Einstein of advocacy because he criticized the Nazi regime for violations of civil liberties. The academy said he should have remained silent, neutral, and objective.

WHAT A MESS. What an ungodly mess. My first talk on scientific advocacy was in 1994. Over twenty years ago. It was a fine talk, if I may say so. I thought it answered most of the questions. But the debate has continued and intensified. How should a scientist handle the so-called conflict between scientific objectivity and public responsibility? What is the work of a scientist in a time of impending climate chaos and environmental disruption? The discussion endlessly repeats. I'm sick of it.

You know what I think? I think we're not making any progress on this because we're asking all the wrong questions. They are the wrong questions because they are the old questions. While we have been debating advocacy, the world has changed around us. We're now dealing with projections that if we don't quickly change course, the world may soon be uninhabitable by many of the creatures, including the human creatures, that evolved during the previous period of relative climate stability. The world is desperate for change. We have one chance to do this right. We have, maybe, one decade to get it done. We need all hands on deck, as they say, and that includes scientists. But most scientists are down in the hold, muzzled by the vague but real fear that if they speak out, they will be punished for "advocacy," the cardinal sin of science.

Let's change the questions and see what progress can be made.

Scientists ask, "What is advocacy?" They should ask, "What damage is done (or whose interests are served) by the chronic ambiguity of the term advocacy*"?*

I don't know what advocacy is. When I ask scientists, I get as many answers as people I ask. When I ask the university administrators, who police the activities of the scientists, I get vague and inconsistent responses. When I ask people whose work has been criticized for advocacy, I get a cynical response: "Advocacy is coming to a conclusion different from the one that serves the university's economic or political interests."

Let's try an experiment. Say you are a scientist and you have come up with a simple cure for a terrible, fatal disease. Call it whatever you want. Call it the dreaded *morbus*. Which of the following statements is advocacy, inappropriate (maybe unprofessional) for a scientist to say?

1. Morbus causes terrible suffering, especially among children—especially among poor children.
2. I will tell you what it's like to die of morbus. First, you lose the ability to hear, then to see; then you are cut off from all sensation except pain.
3. Using this simple cure, we can wipe morbus off the face of the Earth.
4. Sometimes I wake up at night thinking about the suffering of children. Sometimes I wake up crying.
5. It's wrong to let children suffer when we have an easy cure for their pain.
6. Senators, we should publically fund a program that increases funding for morbus research.
7. Senators, we should publically fund a program that eradicates morbus around the world.
8. The world would be a better place without morbus.

Follow the options: The first is a statement of fact, based on scientific evidence. The second is a scientific description of the disease. The third is still science, an evidence-based, falsifiable prediction. No problems so far, right? The fourth is an honest description of the scientist's emotions. Oops: Is it okay for a scientist to talk about his emotions? The fifth is a normative judgment, a moral affirmation: It's wrong to let children suffer. May a scientist say this? If not, why not? Must she say it? If not, why not? The sixth advocates a public policy in favor of science. That's okay, right? It's always okay for a scientist to advocate for funding. The seventh advocates for a public policy in service to world health. Any problems with this? No? But it's frankly advocating! The last is a value statement, an affirmation of what is good. Scientists aren't allowed to say that, right? At least not *qua* scientists. Maybe if they take off their science hats?

As I say, it's a mess. And this is an easy example. Let us imagine that it's not curing a child's disease that is at issue but, say, prohibiting the injection of poisonous fracking chemicals into a community's water supply. Suddenly, even speaking the truth about the subject is attacked as advocacy, or "far-left environmental advocacy," as the Heartland Institute calls it.

It really matters that no one knows what advocacy is, although everyone knows that scientists shouldn't do it (with their scientific hats on). What damage does this do?

The ambiguity of advocacy widens the area of caution. If the campground host tells you, "Don't go where there are snakes," and you have no idea where the snakes are, will you walk any of the trails? In the face of uncertainty, won't prudence keep you in your tent?

Or consider Mark Twain's cat. "A cat who sits on a hot stove will never sit on a hot stove again," Twain wrote. "But he won't sit on a cold stove, either." The cat has no idea which stovetops are hot and which are cold, but the smart cat stays away entirely. And this is exactly the effect on scientists of uncertain and changing standards of scientific conduct regarding advocating in the public sphere.

If I were CEO of a destructive industry whose profits depended on lies and concealments, and if my power to tell lies was threatened by science (which is arguably the greatest system of truth and progress ever in the history of the solar system), what would I do? I would have to find a way to silence the entire institution of science. But I wouldn't do it myself. I'd find a way to make scientists silence themselves. *They're a bunch of sitting ducks,* I would chortle. After all, scientists as a whole have an overriding faith in empirical evidence as the sole arbiter of truth. Consequently, they tend to distrust emotions and misunderstand moral and political judgments; anything that cannot be justified by empirical evidence alone will sully the work, many scientists believe. So it's going to be easy to turn their machines of peer review and tenure decisions toward the work of handcuffing themselves to the amoral, unemotional, unearthly world of experimental models. *Heh heh heh.* The silencing is well begun.

And then to make this even more diabolical, I would prohibit any clean definition of what scientists may and may not do. The silencing is complete, except for some mumbling and a few outlaw scientists who can be discredited or destroyed.

I have heard it said that the charge of advocacy is a sword that scientists carefully hone and hand to their enemies to disembowel them. It's not a sword. It's a disease-infected club. Nothing neat about it. The chronically blurred boundaries of the forbidden realm of advocacy serve the interests of silencing, and so they serve the interests of those whose profits depend on hiding the truth.

Scientists ask, "Will advocacy damage my credibility?" They should ask, "Why should anybody believe me if I don't act in ways that are consistent with my claims?"

Five hundred and twenty scientists from forty-four countries, led by Stanford scientists, signed a document making a notable prediction: "By the time today's children reach middle age, it is extremely likely that Earth's life-support systems, critical for human prosperity and existence, will be irretrievably damaged by the magnitude, global extent, and combination of these human-caused environmental stressors, unless we take concrete, immediate actions to ensure a sustainable, high-quality future." They handed the document to California governor Jerry Brown and went back to their offices.

This was an important, courageous, and eloquent act. Did people take the letter seriously? Not especially. Why not? I would guess it's because not even the scientists took their letter seriously. If they really believed what they wrote—that by the time their children were middle-aged, the life-support systems of the Earth would be irretrievably damaged—they wouldn't be writing a letter. They wouldn't be tossing balls with those same beloved children. They would be out on the street, demanding action. Where is the proof of the scientific claims? It's not in the data alone but also in the acts of those who report the data. A climate scientist who reports the world-disrupting effects of carbon dioxide pollution to his peers, says nothing to the public, and then jets off to the next meeting in Sweden has the credibility of a surgeon general who smokes Lucky Strikes.

The Quaker civil rights leader Bayard Rustin said: "The proof that one truly believes is in action."

One might think that we've got two issues. Scientific integrity. Moral integrity. I believe that they are the same thing, and they both mean wholeness, a matching between what we believe is right and good and true, and how we act. I believe that on issues such as global warming, species extinction, pollution of water and soil, and acidification of the ocean—all with their immense implications—it is an abdication of one's responsibility as a scientist, entrusted with the truth, to fail to speak out in ways that are educational and that prompt healthy societal change.

Consider again the child in the river. Let us say you, a scientist, are walking along the river and you see a child struggling in the current. Let us say that you remark to an onlooker, "Look, this is a fact: If no one helps that child, she will drown." Is that the end of your obligation? What if the onlooker doesn't jump in? Is it okay to sign a letter with 520 scientists, saying, "Look, if no one helps that child, she will drown"? Is it okay to testify to Congress, saying, "Look, if no one helps that child, she will drown"? Is it okay to tear off your shoes and jump in? Multiple-choice quiz: Which of these actions damages your credibility as a reporter of the facts?

Notice that it's integrity, this wholeness, which makes the old citizen's hat versus scientist's hat distinction absolutely impossible. A scientist is a moral being and a citizen. It is the scientist's very knowledge, as a scientist, that makes her voice powerful. Why, on issues that shake the Earth, would anyone give away that power?

I have a dream that one day, all of my university's hundreds of scientists will gather in the streets in their lab coats to tell the Oregon State University Foundation that it must sell its investments in a fossil-fuel-soaked future. I dream that one day, all of the health scientists and hospital physicians, with blood still on their scrubs, will stand on the state capitol steps and tell the state legislators that—in the name of human health—they have to put a price on carbon. That one day, every scientist who studies a living thing, a frog or a beetle or a subterranean fungus, will carry that living thing (in a jar, if necessary) in a long trudge from campus to the city council chambers to tell city council that it cannot, it may not, drain one more marsh or fill one more wetland for one more Walmart parking lot. Nor can it allow anyone to fell one more intact forest community. I will help by designing the scientist's hat, which they all can wear.

Bob Dylan said it straight:

What good am I if I know and don't do
If I see and don't say, if I look right through you
If I turn a deaf ear to the thunderin' sky
What good am I?

Whether or not to take action is not the issue. Not taking action *is* taking action—in favor of the dominant paradigm, in favor of the present course of history. It is impossible to sit this one out when sitting it out is a political statement.

Scientists ask, "Will advocacy damage my objectivity?" They should ask, "What damage do we do by pretending that science is value-neutral?"

First, let's be clear about what philosophers of science have been telling us for a century: that despite claims to the contrary, science is not free of values. "Science does not stand apart from human intention and purpose. The human endeavor to know the world is guided by our needs and interests; there is no other comprehensible account of science," philosopher Kristin Shrader-Frechette wrote. Normative judgments enter into science as a function of personal or institutional values and subjective decisions—*What shall I study? What questions shall I ask? What questions will I not ask? What counts as confirmation? What will I do with my results?* "How facts are investigated, selected, and interpreted depends upon one's values, which are colored by how one sees the world," Shrader-Frechette concludes.

But here's the important point: The necessary and ubiquitous roles that values play in scientific inquiry don't destroy its objectivity. Why not? Because science has built-in methodologies to control for individual bias and correct mistakes that might be introduced by subjective judgment—criteria built in to experimental design and review. No one claims these self-correcting methodologies are perfect, but they are part of what it means to do science.

It follows that the real question is, "What damage are we doing by pretending that the practice of science is value-neutral?"

Here's one kind of damage: By pretending that values play no part in science, scientists can avoid thinking about, or assessing, the values that shape their work.

For example, I would venture to guess that in the College of Forestry at my university, one normative assumption is that all forests are potentially fungible—can be exchanged for money. Forests fall in the category of commodities, the way children, for example, do not. And so forests may honorably be treated in ways in which children cannot. This is an assumption worth thinking about. But if no one is aware of the power of this presumption, then it will shape the work of the forest scientists, and their work will reinforce and affirm it.

Another damage caused by separating value questions from the work of science is that scientists risk teaching their students moral irresponsibility.

Facts have consequences in the real world. If one teaches, for example, that heavy silt in streams can suffocate fish but does not invite students to draw any conclusions about where forests should and should not be cut, then one is teaching students to disconnect facts and their consequences. That is instruction in moral irresponsibility.

If a scientist thinks that to be objective is to be neutral, then she takes sides without even knowing it.

Here, again, I turn gratefully to the work of Kristin Shrader-Frechette. Objectivity and neutrality are not the same thing, she argues: Not all positions are equally defensible. If they are not, then real objectivity requires one to represent indefensible positions as indefensible and less defensible positions as less defensible. It follows that neutrality on the issue of whether climate change is real—maybe it is, maybe it isn't—is a kind of deception. Objectivity requires an honest statement of the conclusion supported by the evidence. To represent objectivity as neutrality in the face of a great hazard or threat is simply to serve the interests of those responsible for the threat.

Some time ago, twenty prominent ecologists signed a letter to the journal *Science*, saying, among other things, that students in ecology should contribute to "stemming the tide of environmental degradation and the associated losses of biodiversity and ecological services. . . . Much of what we study is fast disappearing. . . .

Ecologists have a responsibility to humanity, one that we are not yet discharging adequately."

In a printed response, another scientist wrote, "I thought that was a travesty. The public won't know when to trust us." But which person will you trust? Ecologist A, who says, "I study biodiversity and ecological services. It is better that ecosystems thrive than not"? Or ecologist B, who says, "I study biodiversity and ecological services. I take no stand on whether it's better if the plants and animals live or die"?

THE CHOICES SCIENTISTS must make are desperately complicated. But I can speak about myself. I am frightened for the scientists who feel compelled to tell the truth about the terrible entanglement of craven politicians and ruthless resource development and multibillion-dollar global industries. It's a professionally risky choice. But this is a dangerous time. In my mind, scientists who choose this work are the superheroes of this turning point in history. It is important to note that superheroes break the chains that hold them; they do not forge them.

the work of nature writers

Aldo Leopold wrote that one of the costs of an ecological education is that one "lives alone in a world of wounds." He goes on. "Much of the damage inflicted on land is quite invisible to laymen. An ecologist must either harden his shell and make believe that the consequences of science are none of his business, or he must be the doctor who sees the marks of death in a community that believes itself well."

Yes, we live in a world of wounds. But no, we are not alone in noticing. It's not just ecologists who can't help but see the injuries to the landscape. Like so many others who are paying attention to the natural world, nature writers also live in a world of wounds, celebrating frog song and meadowlarks even as they vanish. We can see the gashes in the ecological communities. We can smell the blood that is seeping out from under the parking lot at the mall. We can smell the scorched brakes as we skid to a stop on this uber-Western, uber-capitalist dead-end road. We can hear the crunch of unrestrained capitalism eating its own feet. We can hear the world cry out, all the messages it is sending us in the languages of fire and storm. What are nature writers to do?

This is the question I ask myself over and over again. It isn't an easy question to answer. There is no time to waste. There is no clear route forward. Terry Tempest Williams, author of *Refuge,* told me that she "no longer has the luxury of writing lyric prose." Bill McKibben, author of *Hope, Human and Wild,* has turned entirely to activist narratives. Wendell Berry, poet and author of *Life Is a Miracle,* wrote a full-page petition in the *New York Times.* Janisse Ray, author of *Ecology of a Cracker Childhood,* keeps on writing, but she will not step on a plane to travel for workshops or book tours. Can a writer continue to celebrate what is beautiful on this planet, even as it's paved and poisoned and lost for all time? These are not idle questions. What is our work in this world, at this point in time, when 4.9 billion years of universe history have come down to this moment of decision?

Robin Wall Kimmerer, a writer and botanist with deep roots in Potawatomi culture, says that if you want to know what your responsibilities are, you should ask, *What are my gifts?* The lark has the gift of a beautiful song, so its responsibility is to wake the world each morning. The salmon has the gift of rich red flesh, so its responsibility is to feed the people. The sun has the gift of light, so its responsibility is to show the way. What are our gifts, we writers?

I believe we have four particular gifts. Each one gestures toward what we must do.

One. Like others, writers have the gift of memory, of what remains and what has vanished. Our work, then, is to testify to what is lost. Forgetting is a terrible danger. Ecologists call it the sliding ecological baseline. If people don't remember waking up to birdsong or listening to frogs at night, if the children know only empty or culvert-choked streams, if they know only fouled estuaries and have never seen an uncut forest—if, in fact, they can't distinguish a tree plantation from a forest—then it is possible gradually to come to believe that a stripped-down, dangerous, dammed-up, paved-over, poisoned, bulldozed, radioactive land is the norm: the way it's always been, the way it must always be.

An early settler in Oregon, who watched farmers create muddy stump farms from what had been one of the world's greatest forests, wrote to his family back east: "When the last forest is cut, when the last salmon is caught, when the last river is dammed, we will sit by the river and weep." But it has been much worse than that. Oregonians don't weep. We don't even notice. We cross the bridge over the Willamette River and never see the silver marshes and rafts of ducks and braided channels that once stretched from the coast range to the foothills of the Cascades. Who will hold the vast richness of the planet in humanity's collective mind?

I believe that a sliding ecological baseline helps explain the sliding moral baseline, the degradation of social standards of what is decent and true. Over a hundred years, the settlers replaced an ethic of aspiration (How can we achieve full cultural and ecological thriving?) with an ethic of regulation (How far can I stretch these rules until I'm stopped or fined?). We think of ourselves as good people, but we poison the lawns where children play. We congratulate ourselves on leaving no child behind, while we bulldoze habitats no child will ever see. We install fluorescent lightbulbs and fly off to Belize. We ask so little of ourselves.

And so we are witness also to a sliding baseline of hope, where even our vision of what might be has been reduced, even our vision of human possibility is shrunken and cynical, and even our children's imaginations are stripped, clearcut, and impoverished.

Hold on to the memories of a free-flowing stream. Write about blackbirds in wheeling flocks over marshlands as deep as the horizon. In a boat rocking on a turquoise sea, write about the confetti of fish, a sea flashing with angels' fins and narrow teeth and bulging eyeballs. Humanity will need to remember these, or we will not know how to measure our terrible loss.

Two. Like others, writers have the gift of imagination at a time that calls for the greatest exercise of human imagination the world has ever seen.

Imagine the Great Turning. Imagine the new dreams. Reimagine the ancient story of the Earth—one beautiful, mysterious, interdependent system, deeply worthy. Reimagine the story of human nature to replace the grotesquerie of capitalism's portrait of radically selfish, eternally unsatisfied consumers; remind humanity that we are born with the capacity to love and that we find our greatest flourishing in communities of caring. Reimagine what it means to live a good life, a question older than the Greeks, now being answered almost exclusively by advertisers.

We have to imagine how to live in this world. Then we have to imagine that life into existence in all its particulars. Write the stories about how to live in a world of intricate, absolute interdependence—people and nature, past and future, near neighborhoods and mountain peaks. Write the stories of intimacy with the land, the ecology of love. That is the work of the moral imagination.

Three. Writers have the gift of a sense of wonder, a fragile and ephemeral gift that comes with us into the world as our birthright. Alone among the animals, I suspect, human beings can be astonished. Sometimes I think the most important words a writer can put on paper are *Look, just look, look at this morning, at this damp light, at this wet sidewalk as if you've never seen a morning before.* Then the truth is revealed to us: that we cannot take this for granted, that there is something rather than nothing, and that it is so beautiful. This profound seeing with a poem's powerful lens, this sense of wonder at the world's endless astonishments, this impulse to honor the sacred world, has radical moral consequences. It closes the distance between what is and what ought to be.

The work of the writer is to reawaken a sense of wonder that impels us to act respectfully in the world. There is worth in these products of time and rock and water, far beyond their usefulness to human purposes. The sweep of time, the operations of chance, have created something that leaves us breathless and rejoicing, struggling to understand the very fact of it, its squeaks and songs. It deserves respect—which is to say that a sense of wonder leads us to celebrate and honor the Earth. Our work is to write these stories.

Four. To a certain extent, the writer has the gift of a megaphone, an amplifying factor, and this imposes an obligation to speak the truth in every way a writer knows how. That we live in a world of wounds cannot be denied, but it *is* denied all over the place. Tell people: Trees are dying at twice the rate they used to, at every elevation, trees of different sizes and species. Tell people: One million species of fish, invertebrates, and algae depend on the world's reefs, and the reefs are dying. Tell people: The cost of shutting down coal plants is more than made up by the savings that the closings would create in healthcare costs. Tell people: More than three million people die each year from breathing polluted air. Oh, so many truths to be told.

WRITE AS IF your reader were dying, Annie Dillard advised: "What could you say to a dying person that would not enrage by its triviality?" Now we must write as if the planet were dying. What would you say to a planet in a spasm of extinction? What would you say to those who are paying the costs of climate change in the currency of death? Surely, in a world dangerously slipping away, we need courageously and honestly to ask again the questions every author asks: Who is my audience—now, today, in this world? What is my purpose?

Some types of writing are morally impossible in a state of emergency: anything written solely for self-advancement. Any solipsistic project. Anything intended to stand in the way of truth-telling. Anything, in short, that isn't the most significant use of a writer's life and talents. Otherwise, how could it ever be forgiven by the ones who follow us, who will expect us finally to have escaped the narrow self-interest of our economy and our age?

Some forms of writing will go away. Other new forms will grow from the need for them. Perhaps some of these:

The drum-head pamphlet. Like Thomas Paine, writing on the head of a Revolutionary War drum, lay it out. Lay out the reasons why extractive cultures must change their ways. Lay out the reasons that inspire the activists. Lay out the reasons that expose and shame the politicians. Lay out the reasons that are a template for decision-makers.

The brokenhearted hallelujah. Like Leonard Cohen, singing of loss and love, make clear the beauty of what we stand to lose or what we have already destroyed. Celebrate the microscopic sea angels. Celebrate the children who live in the cold doorways and shanty camps. Celebrate the swamp at the end of the road. Leave no doubt of the magnitude of their value and the enormity of the crime, to let them pass away unnoticed. These are elegies, these are praise songs, these are love stories.

The witness. Like Cassandra howling at the gates of Troy, bear witness to what you know to be true. Tell the truths that have been bent by skilled advertising. Tell the truths that have been concealed by adroit regulations. Tell the truths that have been denied by fear or complacency. Go to the tar fields, go to the broken pipelines. Tell that story. Be the noisy gong and clanging cymbals, and be the love.

The narrative of the moral imagination. With stories and novels and poems, take the reader inside the minds and hearts of those who live the consequences of global warming. Who are they? How do they live? What consoles them? Powerful stories teach empathy, build the power to imagine oneself in another's place, to feel others' sorrow, and so take readers outside the self-absorption that allows the destruction to continue.

The radical imaginary. Reimagine the world. Push out the boundaries of the human imagination, too long hog-tied by mass media, to create the open space where new ideas can flourish. Maybe it is easier to imagine the end of the world than the end of capitalism or fossil fuels or terminal selfishness. But this is the work that calls us to imagine new lifeways into existence. Writers may not be able to save the old world, but they can help create the new one.

The indictment. Like Jefferson listing the repeated injuries and usurpations, let facts be submitted to a candid world. This is the literature of outrage. How did we come to embrace an economic system that would wreck the world? What iniquity allows it to continue?

The apologia. Finally this: Write to the future. Try to explain how we could allow the devastation of the world, how we could leave those who follow us only an impoverished, stripped, and dangerously unstable time. Ask their forgiveness. This is the literature of prayer. Is it possible to write on your knees, weeping?

THERE'S ANOTHER FORM of writing that might be helpful: ghostwriting. Activists and change-makers of all kinds are overwhelmed with the work to be done, and sometimes they are at a loss for words. Writers can help them by drafting templates for a variety of purposes.

Here is a template for the letter a university president can send to the board of directors of her university foundation, asking them to divest from fossil fuels. This is Form Letter 400, which is the parts per million of greenhouse gases currently in the atmosphere. Form Letter 450 will have to be a lot tougher.

FORM LETTER 400: DIVESTING FROM FOSSIL FUELS

To the community of [your university's name here] and all those who believe in its moral integrity,

Today, I am asking the university's [foundation/board of trustees] to end all investment in fossil fuel industries.

[Your university's name here] is deeply committed to the thriving of the Earth and all its communities, ecological and cultural. Across campus and around the world, our students, faculty, and alumni are working to create understanding that promotes human flourishing and strengthens the Earth systems that sustain us.

[Choose appropriate examples: A paleoclimatologist probes the ocean sediments for evidence of global change / An ethics class studies questions of justice and compassion raised by climate change / An entire array of solar panels powers segments of campus / Engineering faculty lead students to innovations in alternative energies / Our writers and artists celebrate the astonishing, nourishing world.]

The university's science and scholarship lead us to understand that the increasing concentrations of greenhouse gases in the atmosphere are causing changes to the climate that threaten to damage beyond repair the systems that sustain our lives and all other lives on Earth. We have come to understand, too, that one of the primary causes of the increase is the burning of fossil fuel in all its forms, including oil, natural gas, and coal. We know, through our own work, that there are alternative energies that are both economically and environmentally sustainable.

Moreover, we understand that damaging Earth's life-support systems violates our moral obligations to our students to leave them, not an impoverished and dangerously unstable world, but a world as rich in possibilities and promise as our own has been. We understand that it is unjust that the costs of burning fossil fuels in this generation will be borne by the blameless and the voiceless. We understand our duties of compassion to those who will be routed from their homes by flood or drought, storm or starvation.

We therefore understand that for the university to continue to profit from investments in fossil fuels not only damages the people we are sworn to serve—the people of the state, the nation, the world, and the young people who are our students—but it does serious damage to the moral integrity of the institution. It is indefensible to earn money by investing in what we ourselves have helped the world understand is life-diminishing and unjust.

Some board members may argue that divestiture violates their fiduciary responsibilities by exposing the funds to increased risk. Research cited in the *Chronicle of Higher Education* and elsewhere demonstrates that this is not the case. But even if it were, the argument is beside the point. There are things that a university does not do to make money, because they are wrong and hurtful. And there are some things it must do, even at the fear of some cost to itself. This is the core of morally and socially responsible behavior.

For [number of years you have been president], we have worked together to ensure the long-term viability of this university. I am grateful for your commitment. This letter is a call to recognize that the greatest threat to the viability of a university is a world in the turmoil of an increasingly violent and dangerously unstable climate. Under those circumstances, responsible fiscal planning will be unimaginably difficult. Doing our part to slow global warming by divesting from fossil fuel industries will allow us honestly to claim that we at [your university's name here] are responsibly and wisely investing in the future of our students, our state, and our planet.

Sincerely,
[your name here], President

❖

AND HERE'S A template for a speech at a climate rally, designed to be shouted over a squealing microphone to a couple hundred dancing people and their dogs. It's called "It's a Bad Day," which is a joke, because of course any day is going to be a great day when hundreds, or tens of thousands, of people rally in the streets for action to stop the carbon catastrophe.

IT'S A BAD DAY FOR REX TILLERSON

This is a bad day for pipelines and export terminals and tankers and coal trains. This is a bad day for the Koch brothers, and Rex Tillerson of ExxonMobil, and anyone else who would trade the life-supporting systems of the Earth for short-term profits.

This is a bad day for the universities that are holding on to their last investments in fossil fuels, insisting on their right to profit from death and extinction, even as their own scientists warn them that fossil fuels will carry the world, smoking and stinking, to the end of life as we know it on this planet.

This is the last day for despair. It is the last day to say it's too late, that there is nothing anyone can do. It is a day to awaken to the fact that we are not helpless at all—that we have the knowledge and the courage and the joyous communities it will take to make the Great Turning away from death and toward a reinvented life.

This is the last day for lies and excuses and delay. It is the last term in office for elected officials who will not or cannot protect the future. It is the last day that anyone can be silent about climate change.

And so, this is a good day for the hoofed and winged things. It's a great day for the small children of all species, a great day for ice and oceans, a great day for reliable rain.

This is a great day for justice, and the right of all beings to clean air, clean energy, healthy food, and freshwater.

This is a great day for sanity and imagination. Imagine a world without wars for oil. Imagine a world without the din and dirt of internal combustion engines. Imagine democracy without the corrupting wealth of coal barons. Imagine a world powered, as plants are powered, by the sun.

Today is the day when everything changes. In every struggle for justice, there is a turning point, a tipping point, when what was unimaginable becomes inevitable. It is the day when the people pour into the street to reclaim their futures, and the future of all the glorious lives on Earth. Life is not a commodity, to be bought and sold, wrecked and ransacked, for the immense but short-term profit of a few sullen and frightened men. The profusion of life is a sacred trust, a great and glorious gift, to be honored and protected, and passed along, intact and singing, to the next generations of all living things.

❖

OKAY. AND HERE'S something I wrote for you, for when you are tired. Feel free to sign it and leave it at the door of a climate activist friend who is too tired to sleep. Maybe, with a ribbon, tie it to a bottle of wine.

AND WHY YOU MUST

No letters to your congressional representatives tonight. No second cup of shade-grown coffee. No Web search for rates of ice cap melting or declining numbers of polar bears.

Turn off the lights. Go outside. Shut the door behind you.

Maybe rain has fallen all evening and the moon, when it emerges between the clouds, glows on the flooded streets and silhouettes leafless maple trees lining the curb. Maybe the tide is low under the docks and warehouses, and the air is briny with kelp. Maybe cold air is sinking off the mountain, following the river-wall into town, bringing smells of snow and damp pines. Starlings roost in a row on the rim of the supermarket, their wet backs blinking red and yellow as neon lights flash behind them. In the gutter, the same lights redden small pressure waves that build and break against crescents of fallen leaves.

Let the reliable rhythms of the moon and the tides reassure you. Let the smells return memories of other seas and times. Let the reflecting light magnify your perception. Let the rhythm of the rushing water flood your spirit. Walk and walk until your heart is full.

Then you will remember why you try so hard to protect this beloved world, and why you must.

the work of wilderness

CITIZENS, SCIENTISTS, WRITERS: We all have our assignments in this work of "saving the world." And what about wilderness? Does it have a role to play in the unfolding drama? Wilderness has taken a lot of hits lately, both conceptual and physical. Even as social critics are arguing that wilderness doesn't exist except as a social construct that humans overlay on the land, extractive industries shove up against its artificial boundaries, ready to poke pipes horizontally under the land, and defenders of the American Way stand armed and ready at the boundaries, alert for wolves. In my Oregon mountains, global warming is reducing the snowpack and the ability of the trees to withstand beetle infestations, and wildfires have blackened every one of my favorite haunts. I decided to return to one of these, to think about what wilderness might mean for the future. What I found there, I will call the *new* geography of hope.

IN THE BURNT RUBBLE OF THE GEOGRAPHY OF HOPE

After some serious wandering through this thicket of fire-blackened spars, Frank and I were able to find the burnt butt of the Mount

Jefferson Wilderness boundary sign and a vague indentation where the trail had been. Some of the great Douglas firs still stood, but their limbs had burned to stubs and their bark hung in scabby patches on the trunks. The wildfire had burned entirely through the smaller trees, the hemlocks and firs, and toppled them in tangled heaps. I know that forests burn—that's the way of forests—but this fire didn't just burn the trees to the ground. It burned the ground itself and didn't stop there, but burned even the roots in the ground, hollowing out a warren of tunnels and caves. My husband and I stepped along the shadow of the trail, careful as cats, but even so we broke through more often than we thought was safe, sinking to a shin or knee in an emptiness where roots had been. Our boots were dusted white as bones. And even worse: Expecting the usual silence of the wilderness, we were not prepared for the racket of a billion beetles, gnawing dead trees.

We were as gloomy as the day, depressed by the deep irony of walking through the charred stubble of a so-called geography of hope. That's how Wallace Stegner described wilderness fifty years ago. Designated wilderness areas, he wrote, are a "means of reassuring ourselves of our sanity as creatures, a part of the geography of hope." Exactly so. I too believed that the industrialized growth economy was certifiably insane: obsessed with pointless, greedy getting, unable to conform its behavior to standards of right and wrong. And I knew the hope in wildness. I had stood above tree line on Mount Jefferson in Oregon, where snowcapped volcanoes parade the length of the Cascade Range, and I had felt the limitless possibilities and endless expanses that seemed unconstrained even by gravity—the weightlessness of open space. In that wilderness, even avalanche lilies unfurling at the very edge of snowfields testified to deep evolutionary wisdom and time without end. What could be a better marker of sanity than to value and protect places like this?

But dear god, wild places are really taking it in the chin now, as global warming strengthens its grip on the land. Because wilderness designation has protected the most vulnerable places, it's no surprise that these are being hit hard and first: the high airy mountaintops, the steep forests, the coastal marshes and arctic edges, the last refuges of specialized species, places of high drama and deeply felt significance. From the wilderness mountaintop, the "end of nature" can now be clearly seen. This decade has become a race between the end of the fossil fuel economy and the end of the Eocene era. Where is the hope in wilderness now, in the polar bear and rotten ice that have become the face of climate catastrophe? Where is the hope in the orange-needled, beetle-killed forests, facing into a hot wind? Where is the hope in a shrunken, dusty glacier at the base of a raw couloir?

I am struggling to know what to make of the geography of hope now, in a time of drought and storm.

At the top of the next hill, we stopped to dig out our water bottles. The incessant assault of beetles' buzz was driving me mad, and so was the gray and ashen land that seemed to stutter like a silent film as we walked past the framing spars. But here was a flash of movement between the trees. We heard sharp raps, repeated, slowly at first, then gaining speed. Another flick of wings, and a black-backed woodpecker attached itself to the spar I was leaning against. The bird began again to drum. In the hollow of my chest, I could feel the percussion. Bark flakes fell on the brim of my baseball cap. My husband laughed, which sent the woodpecker sidestepping to the other side of the trunk.

THERE IS NOTHING to do but to reimagine the geography of hope, to think in new ways about the reasons why wilderness designation is profoundly sane on a planet reeling under a pathology of greed and shortsightedness. I want to say that the wounded wilderness landscape offers a new kind of hope. It's a defiant hope, a ferocious hope, sometimes a desperate hope, and maybe a redemptive hope, a hope that as plants and animals change their very lifeways in a brutal race against climate change, humans also can change their lifeways, reinventing what it means to be human in a finite, deeply interdependent, and generously beautiful world.

A defiant hope. Like a fence, the boundaries of a wilderness area contain a landscape. But, more importantly, they exclude a landscape. A wilderness is testimony to the human will to say, *No, the industrial growth economy will not cross this line.* Reckless disdain for the natural world has no place here. Here, the landscape is valued in itself, for its own sake, not for the profit that might be wrung from it. Fracking pads must stop short of this red rock canyon. Oil wells must stop short of this arctic mountain range. Water-sucking mines must stop short of this desert spring. Cattle must find a different place to blow and belch. A wilderness echoes with moral outrage: There are business plans that are hideous and cruel, and they will not do business here.

A ferocious hope. But even if wilderness legislation can bar the entry of extractive industry, nothing can protect any area from the effects of global warming. No wilderness permit system can exclude the storms. No federal agency can call rain into a fearsomely parched land. No boundary check-in station can control the tides. The only way to protect the wilderness from global warming is to take ferocious action against the causes of global warming itself. That means that anyone who loves the wilderness is called to action against the fossil fuel industries, which have shown themselves perfectly willing to let destruction of wild places be one of the costs of doing business.

A desperate hope. There is no doubt about the geophysical worth of wilderness in a time of climate change. The desperate hope is that healthy forests and soils can sequester carbon as fast as humankind can pour carbon dioxide into the air. Obviously, the more healthy ecosystems, the better the chance. So the more intact wilderness, tangled banks, heavily breathing forests, greening jungles, tundra, and dense black soil are present on the planet, the more carbon dioxide they will suck from the air. To the extent that designated wilderness saves intact ecosystems, and so saves carbon sinks, it is the great hope of the reeling world. A sane policy would rapidly expand protected land, not asking, *Is it pristine? Is it untrammeled?* But asking only, *Does it breathe?*

A redemptive hope. There is one place you can go on God's good Earth where you have little choice but to be your best self. That is the wilderness. It is simply against the law to be a greedy, reckless pig in the wilderness. It is simply against the law to steal or vandalize whatever you want. A wilderness sojourner is called to a kind of self-restraint that is rare in life. It's a generosity of spirit that takes only what is given and returns it in gratitude and care. This is a fact of great importance: A week in the wilderness is proof that a human being is capable of being a good and decent citizen of the Earth.

The surprise is that when travelers cross that wilderness boundary, they have the ability to slip from one level of being to another, from being people surrounded by and obsessed with and dependent on multitudinous stuff torn from the Earth, to being people whose greatest pleasures are simplicity and a close connection to something greater than they are, something wiser and more powerful. If there is not hope in this proof of the transmutability of human character, I don't know where it can be found.

I've started to think that I am drawn to the wilderness because I want to free myself to become a person I believe in. I'm tired of doing what I think is wrong. It grieves me. I know full well that my car is emitting carbon dioxide that will create real hardship for my grandchildren. Even as I board the planes, I know that the carbon costs of the cross-country flights I take will be paid by bewildered children. I do it anyway. But a week in the wilderness near my home is a chance to re-create myself, to reshape my life, to express my deepest values, to practice being the person I want to be. The freedom of the wilderness is not that I can do anything I want; it's the opposite. Here I am free to restrain myself by my own sense of right and wrong, to go AWOL from the industrial growth economy's war against the world.

On the back of a fallen spar, we teetered across a draw, then climbed a small bald. From here, bare and bristled hills rolled over to gray clouds that obscured the high peaks. In that gray land, I was glad for the red fabric of my husband's parka. I followed him around a palisade of blackened spars propped against a cliff. In gusts of ash, he traipsed over the rise and stopped dead. I was not prepared for what I saw when I stood beside him: In a low space, water had gathered into a small oval pond that was vivid green, and all the greener for its bed of ash. We hurried toward it—who wouldn't hurry toward what is green and growing? Green algae bubbled in a broad band around the edge of the pond and spread in green cirrus clouds from its center. Black insects peppered the green billows. So unexpected, so lavishly fecund, the pond could have bubbled up from the welling springs of life itself. We sat beside it. After a time, we pulled out cheese sandwiches and ate them, picnicking by a pea green proto-lake in the cinders.

In times that seem grim and rootless, when even the ground gives way under my feet, I will enter a new geography of hope. I have loved this wilderness once, and I will love it still. The dark, ferny-kneed forest and shy owls, the soft trails, the smell of pine and bracken are gone—maybe gone forever from a sizzling hot future. I don't know how to bear the dead weight of this sorrow and of this shame. I do know that what remains is a wilderness of sinewy, raw-boned possibility. Just that. Possibility, the creative urgency of life unfurling in the dark folds of the land, the fertility of the human imagination, and the expansive embrace of the human heart. Here is the home of hope.

PART IV

a call to act

really hard questions

BEHIND THE SHARK tank at the aquarium, next to the coffee machine in the church basement, in the mosquito-plagued shade under a big oak, behind the gas jets in the chemistry classroom, in blazing sun in the middle of the street, in the bookstore aisle next to HORROR, in front of a display of Elvis portraits on velvet—I swear to god, I have spoken about climate ethics in every possible place. I love the people who come, full of wisdom, desperately caring, oddly joyous. Invariably, the very best part of the event grows from the intake of breath that comes after "Thank you. Are there any questions?" There are. There always are good, important questions, ethical questions, and they come up repeatedly.

They terrify me. First, because they are tough; they raise some of the hardest philosophical puzzles about the human condition. Second, because they are paralyzing; until we can address these questions, our culture will not make any significant progress on turning away from the carbon catastrophe. Third, because they are real; they are cries from the hearts of people struggling to keep their hearts from breaking.

So here are the three hardest questions, and the best I can do to answer them.

we have met the enemy, and is he us?

Pogo and Porkypine are picking their way under the overhanging limbs of a swampy forest. "Ah, Pogo," says Porkypine, "the beauty of the forest primeval gets me in the heart."

"It gets me in the feet, Porkypine," Pogo says.

The frame pulls back and reveals that the forest primeval is a dump, packed with rusted bedframes, a broken bathtub, a sprung tire, cracked bottles, industrial sludge.

"It *is* hard walkin' on this stuff," Porkypine acknowledges, sitting down on a root.

"Yep, son," says Pogo. "We have met the enemy and he is us."

Walt Kelly published his cartoon on Earth Day 1970, more than forty-five years ago, and it still comes up in the question periods after, I think I can honestly say, every single one of my public presentations. In large part, it is a question from people of conscience who acknowledge their part in the fossil fuel economy and are honest enough to accept the moral responsibilities that come with their decisions. Knowing they are not "without sin," they will not "cast the first stone" at the fossil fuel industry. Fair enough.

But in equal part, the question is a challenge to the right of anyone to criticize the fossil fuel industry until they have purified themselves of any taint of oil, plastic, fertilizer, jet fuel. The question implies that because no one can lay claim to this purity, nobody has the right to criticize the fossil fuel industry. This fear of hypocrisy is absolutely immobilizing. Here is Harvard's president, Drew Gilpin Faust, explaining why Harvard must not divest from fossil fuels: "I also find a troubling inconsistency in the notion that, as an investor, we should boycott a whole class of companies at the same time that, as individuals and as a community, we are extensively relying on those companies' products and services."

This is a logical fallacy, President Faust. If you want its Latin name, here it is: *argumentum ad hominem (tu quoque)*—literally, the argument to the man (you're another). Here's an example: "His claim that I should quit excessively drinking is probably false, because he's an alcoholic himself." Oh, for crying out loud: His alcoholism doesn't mean he's wrong about mine. Your reasoning has the same pattern: "The claim that a respected university should not invest in fossil fuels is probably false, because the university burns oil." That is a non sequitur—it doesn't follow. Maybe divestment would reveal the respected university as hypocritical, but even hypocrites can tell the truth.

The truth or falsity of a claim is a matter of the relation between the claim and the real world, not a matter of who makes the claim. Even so, I would consider this ad hominem argument to be the biggest reason for the public's thundering silence about climate change—a public that doesn't ask *why* it is dependent on fossil fuels but rather questions its own moral standing to protest its dependence.

It may be one of the biggest triumphs of Big Oil, to make consumers blame themselves for climate change, even while the corporations are spending billions to transform us into mindless consumers of self-destructive (but cheap) consumer goods and fossil fuels—to make us blame ourselves, even as they leverage their bribes in Congress to be sure that we have no alternative ways to heat our homes or power our tools or travel to work.

So when I hear people say, "We have met the enemy, and he is us," I want to think about this very carefully. Of course, everyone should spend and invest and work and travel more thoughtfully. Of course, everyone should dramatically cut their use of fossil fuels. That said, the fossil fuel industry is very happy to claim that it is simply responding to public demand.

But I didn't demand that corporations cut corners and cause an oil gusher in the Gulf of Mexico. I didn't demand that they undermine the EPA and any other agency that might control fracking under farms. I didn't ask them to lobby/bribe Congress to open oil drilling in the Arctic Ocean. In fact, I didn't ask them to give me unnaturally cheap oil by externalizing all the costs onto me and my children. Rather than responding to demand, the corporations have been manipulating public demand cleverly and pervasively and seemingly without financial or legal constraints.

They build and maintain infrastructures that force consumers to use fossil fuels. They convince politicians to kill or lethally underfund alternative energy or transportation initiatives. They increase demand for energy-intensive products through advertising. They create confusion about the harmful effects of burning fossil fuels. They influence elections to defang regulatory agencies that would limit Big Oil's power to impose risks and costs on others. They do everything they can to be sure that people have no choice but to participate in the oil economy.

It's no surprise that people feel forced to participate in a destructive economy. That's the business plan. Think of the things that are forcing us. One of them is entrenched patterns of transportation that are really entrenched patterns of enriching the oil industry at our expense. Another is entrenched patterns of agriculture and eating that are really entrenched patterns of enriching a few agricultural and seed industries at the expense of our health.

We have met the enemy, and I am going to do everything I can to make sure it isn't me. But while the oil industry is externalizing the costs of pollution and environmental destruction onto me, I will not allow it to externalize its shame.

"ENEMY. BLAME. WHAT is the point of demonizing?" the man in the front row of the balcony asks, and suddenly I'm feeling bad about myself, about my anger, and uncertain about the moral ground I stand on.

We all are trying to do the best we can within the systems that shape our culture and our choices, the argument goes. The corporate leaders of the oil industry are working within a particularly vicious and limiting system. Every quarter, their shareholders examine their financial reports, every three goddamn months, so how can they justify long-range decisions? In all states but one (Oregon), laws require them to make decisions based on the best financial interests of the shareholders, so how can they pass up any chance to maximize profits? The rules that might limit pollution are weak and badly enforced, so how can they be expected to restrain themselves? They have a payroll to meet and it gets more expensive all the time, so how can they justify costly antipollution measures? If you want to blame someone, blame the system. And if you want to blame the system in a democracy, blame yourself. Or so the argument goes.

Makes sense—but hey, wait a minute! What is moral integrity inside a harmful system? It's refusing to play by the rules that you recognize are soul-devouring. It's refusing to play a rigged and dangerous game, even if it means giving up a salary of . . . let's see, after his salary increase last year, $40.3 million for the CEO of ExxonMobil. What is a system, anyway, but a bunch of individuals acting within the system? Which is a choice. A hard choice, maybe, but a choice.

I'm more moved by my Buddhist friends, who advise compassion. Fierce compassion, perhaps, but compassion nonetheless. The Dalai Lama explains, "Destruction of nature and natural resources results from ignorance, greed, and lack of respect for the earth's living things." Not, I infer, from some black seed of evil deep in the soul. I can understand that and generally accept it. But should that keep me from naming the ignorance, the greed, and the lack of respect, which are, the last time I looked, vices? Should compassion for the person prevent me from naming the horror they are creating—not as an unintended by-product of their decisions but as the knowing, directly intended or homicidally reckless, and very carefully calculated consequence of their decisions? And isn't that what culpable wrongdoing is—the knowing and intentional doing of unjustifiable harm?

I believe in moral outrage. I claim it as my right.

Outrage is a measure of what a person cares about the most—what she loves, what ideals she affirms, what breaks her heart or dashes her hopes. I think of it this way: When, eight hundred thousand years ago, a wandering child stepped in a sandy bank alongside what is now the Thames River, she gave scientists what they needed to know the size and shape of her foot, her weight and how she carried it, maybe her species. That particular absence of sand is filled with meaning. In a similar way, the imprint of moral outrage reveals the depth of a person's caring, the beliefs she affirms and how she carries them, maybe even the legacy of belief that has shaped her.

Outrage is a kind of truth-telling, naming wrong for what it is, flinging away excuses. Outrage is a kind of respect, treating powerful decision-makers as responsible human beings, not as victims or helpless pawns or idiots. Outrage is a response that acknowledges, bears witness to, the suffering of innocent beings. Outrage is an expression of love. It is one step away from overpowering grief.

AFTER A WINTER meeting of the Society of Environmental Journalists some years ago, I happened to walk out of the building at exactly the same time as the representative of the American Petroleum Institute. Understand: this was right after the *Deepwater Horizon* oil spill in the Gulf of Mexico, about which, this API representative had just said, and I quote, "There was release that was not prevented."

I hated the oil spill and BP's cynical response. Maybe I hated this representative. He was a huge man—I'd guess six-five. Shaved head. Big black overcoat reaching below his knees. Big black dress shoes with rubber soles. Nevertheless, we were walking side by side into the cold, both of us turning up our collars against the wind. *This is my chance,* I said to myself, *to relate to an oilman in a personal way, and perhaps even learn a little about his heart.*

"So," I said, "Do you have children?"

He knew where I was going with that. He turned to face me straight on. "Don't you ever," he said, "ever. Ever. Ever." He paused. "Ever underestimate the power of the fossil fuel industry."

"Okay," I said. "I guess I won't."

By then, we were crossing a pedestrian bridge over the Clark Fork River. A thick layer of ice coated the bridge, and I was wearing high-heeled shoes. What I really wanted to do was hold this man's hand; he was solid on those big shoes. What I really wanted was to ask him to keep me from sliding on my bum. I should have, but I didn't. Thinking about it now, I am certain he would have helped me. It's fascinating, how complicated is the human relation to one another, and to right and wrong.

HAVING SAID ALL this, having laid the blame for the carbon catastrophe at the big feet of Big Oil, having insisted on the right to lay blame, I have to admit that none of us, you or I, get off the hook. The complication is that none of my individual acts—say I drive to the grocery when I could walk; say I buy a pineapple from Hawaii instead of an apple from Oregon—are noticeably harmful in and of themselves. It's hard to say that I'm *intending* harm; surely I'm not. It's hard to say that I'm *doing* much harm; in an intriguing analysis, climate and energy writer David Roberts does the math and calculates that the average American family each year is responsible for one 13.75 billionth of the increase in excess greenhouse gases. I can't vouch for that exact figure; the point is that none of us is doing much harm, but in the aggregate, over time, around the world, individual acts taken all together are empowering the carbon industries' destruction of the Earth. How can we puzzle through moral responsibility in a case like this?

Philosophers have wrestled with the notion of collective responsibility or aggregate harm for centuries, often after or in the face of terrible events. Was the Holocaust caused by a small cadre of evil men, or was it the result of the studied silence and/or active cooperation of 69,314,000 Germans? If Earth is allowed to bake in the aggregated greenhouse gases of 7.125 billion people over decades, what can be said about moral responsibility?

Quite a lot, actually. If one measures the morality of an act by the benefit or harm it creates—the same old consequentialist calculation—then maybe each individual's thoughtless use of fossil fuels is a vanishingly small wrong. But this runs counter to our moral intuitions, and for good reason. I venture to guess that each of us feels disproportionately bad about small wrongs—failing to return the extra pennies a grocery clerk mistakenly gave me, accidentally stepping on a neighbor's marigold, forgetting to send a birthday card to a demented aunt. Even if (*if*) these small acts do no measurable harm to the world, we know that these do in fact cause measurable harm to *our sense of ourselves as moral people*. The virtue theory of ethics validates the regret we feel when we do what we know is wrong, even if we know it makes no difference at all. Our acts shape who we are as moral beings; and who we are as moral beings shapes our acts. So people are quite right to feel some shame at their role in climate change. I suspect this is one reason why no one really wants to talk about it. It is a shameful subject.

But there's another reason to be concerned about the small contribution each person makes to climate change. Yes, there is the *aggregate* effect of all our acts to consider—your carbon footprint, added to my carbon footprint, and so on to planetary ruin. But it's an unusual act that has merely an aggregate effect. When it comes to climate change, our actions often have *exponential* effects—positive or negative. Why? Because the single most effective way to change a person's actions is to show him that a person he respects has made that change. If I bring a shiny new SUV into my driveway, the effect can be calculated as the increase in my carbon footprint, multiplied by whatever my neighbors are moved to do. In the same way, if I install solar panels on my roof, the effect is caused not only by my act but by my example.

Powerful analogues argue that individual acts do not merely *add up* but increase exponentially, like a fire that creates the wind that fans it. The math is not like this: one racist joke plus another racist joke. The math is a curve that escalates into a culture in which racism in the norm, the usual way of life. I'm afraid that sexual violence is like this. I'm afraid that gun violence is like this. Climate change itself will operate this way, scientists suggest, such that crossing a threshold puts into play factors that will increase and accelerate the release of greenhouse gases—the melting permafrost, the burning forests we have come to know so well.

What lifts my spirits is knowing that positive change works the same exponential way. If the Earth makes the Great Turning, it will not be because of one xeriscaped front yard plus one wind turbine plus one local farm plus one redesigned cookstove, and on and on. It will be because imagination creates more imagination, good creates more good, respect for the land creates more respect, in a swirling whirlwind of change that sweeps away business-as-usual and upends the culture of reckless exploitation. Even as we approach a tipping point in runaway climate change, we are approaching a tipping point in the human conscience. What has been stony denial or astonishing indifference has become a worldwide movement for climate justice and ecological wholeness. We have met the enemy, and he may or may not be us. But we have also met the change-makers, and they definitely are us.

LET'S LOOK AT this cartoon again. Pogo and Porkypine are picking their way under the overhanging limbs of a swampy forest. "Ah, Pogo," says Porkypine, "the beauty of the forest primeval gets me in the heart."

"It gets me in the feet, Porkypine," Pogo says.

The frame pulls back and reveals that the forest primeval is a dump, packed with rusted bedframes, a broken bathtub, a sprung tire, cracked bottles, industrial sludge.

"It is hard walkin' on this stuff," Porkypine acknowledges, sitting down on a root.

Imagine, then—

"Yep, son," says Pogo. He leans over and starts picking up the trash. For his part, Porkypine waddles to town to place his bristling spines into the tires of the pickups loaded with trash to dump in the forest primeval. On Mondays, he visits the halls of power, where he repeatedly pokes his elected representatives in their argyle socks. Wherever the agents of destruction look, there is that damned porcupine with the bent-up stovepipe hat.

what can one person do?

CITY HALL, SAN Jose, big auditorium. The second question came from a young woman with curly red hair piled on her head like ribbon on a present. She stood up and took a breath so deep that it raised her shoulders. This is what she said: "Okay. I'm on board about climate change. I get it. But what can one person do?"

The room murmured. *Mmmm*, the whole auditorium assenting. Heads nodded: white heads, dark heads, heads with hats. This was a very, very good question. I was pretty sure that everybody in the room had already checked off everything on the list of "fifty things you can do to save the planet." These lists are all over the Internet. *Lightbulb lists*, Michael Nelson and I call them, because number one is always "Switch to compact fluorescent lightbulbs." Every single person should, of course, do every one of these things. Yes, you should buy local food. Yes, you should avoid beef, a hideous methane machine. Yes, you should unplug your appliances. Yes, you should refuse to invest in companies that profit from death. Yes, you should vote. Yes, you should take the bus or bike. Yes and yes and yes and yes. No excuses, no delay, no exceptions.

Fine. But the room was asking for something different, something bigger. And frankly, I think they were looking for something that wasn't so . . . well, so lonely and sad.

What can one person do? I say, *Stop being one person.*

Join up with other people, or join a local group, and help save the future by sharing whatever gifts the good Earth has given you: time, money, organizing experience, the ability to lick a stamp or push SEND, a strong heart and a ferocious determination that the world will not go down on your watch.

It's lonely and brittle to live a life you don't believe in. But to be an active part of a purposeful community of caring, to do good alongside other people, to preserve what is lively and beautiful, to really *live* your love for your grandchildren and the struggling animals, to become part of a movement for change? There's joy in that—joy to be in alliance with people who care about the living things you care about more than anything else in the world.

To be sure people are paying attention, I sometimes give a pop quiz. Old professors have a hard time breaking long-standing bad habits.

1. Who said, "Nothing is going to change. The big guys have all the power and all the money. No point in even trying. I can't even imagine how the world might be different."

a. Martin Luther King

b. Gandhi

c. 411,000 demonstrators in New York City for the People's Climate March

d. All of the above

e. None of the above

2. Who said, "I'm only one person. Nothing I can do will make any difference, so why should I put myself out? Besides, I'm pretty busy already."

a. Rosa Parks
b. Rachel Carson
c. Elizabeth Cady Stanton
d. All of the above.
e. None of the above.

3. Who said, "Climate change is the greatest crisis of our time."

a. Barack Obama
b. Pope John Paul II, Ecumenical Patriarch Bartholomew I, and Pope Francis
c. The Dalai Lama
d. All of the above
e. None of the above

Greatest crisis of our time? Maybe greatest crisis of all time. We have a crisis, and we don't need the president or the pope to tell us. A crisis of cosmic proportions. A crisis akin to the crisis that faced the dinosaurs—the only difference being that dinosaurs had brains the size of walnuts, and we have bulging brains and hearts like pumping fists, the ability to see what's coming down the pike, and the ability to imagine that we can do things differently. It doesn't have to be this way. We can absolutely do this.

We don't have to be Martin Luther King, but we have to find his capacity to dream, to envision a different world: a world without wars for oil, a cleaner quieter greener world, a world where how much coal dust and toluene you breathe does not depend on the color of your skin and the decisions of a fat-cat oil executive.

We don't have to be Rachel Carson, but we have to find her raw courage and her stubborn refusal to shut the hell up, like a proper lady would do.

This is the time. This is the top of the hill, when the beautiful blue marble can roll either way. Oil companies are hanging out, waiting to see which way it goes. Politicians (even governors) are hemming and hawing incoherently, hedging their bets, waiting to see which way it goes. Universities are stalling about divestment, waiting to see if anybody makes them do the right thing. Secretary of State John Kerry begs the American people, *Make us do the right thing*. The world is holding its breath. This is what a tipping point looks like. We're there.

What are we waiting for? I encounter a lot of people who say that we're going to need one big disaster. So we have Katrina. So we have the BP *Deepwater Horizon* oil gusher. So we have the biggest fire season in history. So we have massive die-off of forests. And people keep saying, *We are waiting for a big storm to convince us to act*. Or, *We're waiting for a leader*. So we had Al Gore, but people laughed at him so they didn't have to listen. So we have Barack Obama, but Republicans try to block his every move. It's going to have to be concerned people all over the world, creating communities of caring, raising their voices for a just future.

So now what are we going to do?

So many long nights, I've thought about that question. Worse, being a philosopher, I've struggled with the question of how to even *think about* thinking about what we ought to do. Here's the story of one such restless night, deep in Denali National Park on the east fork of the Toklat River, when the river offered me a new way to think about climate action.

THE RULES OF RIVERS

I was clearly not going to sleep that night. It wasn't the heat or the mosquitoes or even the fear of bears that kept me up so much as this constant damn upstream swimming against fury and despair. The cabin was holding its heat, as it was built to do a century ago, but the night air through the screen door was fresh. I pulled on pants and a shirt and walked down to the river.

The Toklat is a shallow river that braids across a good half mile of gravel beds, dried stream courses, and deep-dug channels. Sloshing with meltwater, it clattered along through islands and willow thickets. Banging rocks on cobblestones, surging into confused swells, the gray currents looked unpredictable and chaotic in the late light. Upstream, a chunk of bank fell off and lurched downstream, dirt, roots, and all. The thud of boulders, the scour of silt on stone, the jangle of a wave breaking on every single rock in that river—how many decibels was it? As many as heavy traffic, I was sure. Eighty? Ninety? How fast was it running? Faster than I could ever run. I started to sit on the undercut bank, then thought better of it and moved back into the aspens.

I don't know how to stop a river. I don't know how you can stop anything that has this power: the combined forces of huge money, huge oil, huge greed, huge mistakes dug into the very structure of the land, a tangled braid of paid-off politicians, oblivious consumers, reckless corporations—everybody in some odd way feeling swept along. I picked up a stone and lobbed it into the current. The river gulped it down.

On a spruce bench gnawed by bears, I sat and watched the river churn. It rose as I watched. An arm of the river wrinkled sideways over a bed of cobbles and filled another channel. In a rush of stones, the channel dug itself wider, washing cobbles from the bank and rolling them downstream. The current sloshed into a set of standing waves, then smoothed against a sandy shore. Everywhere there were islands forming in a clattering process of removal and deposit. If I could understand a river, maybe I could understand how to stop it. But this one was complicated, seemingly chaotic, and as opaque as molten bronze.

But there are patterns. The summer before, I had walked along an Oregon river with Gordon Grant, a physical hydrologist who knew the rules of rivers. He taught me about the processes that he liked to call the "ghostly mathematics." The processes of a river are manifestations of energy, he said. A fast, high-energy river will carry particles; the faster the river, the bigger the particles. But when it loses energy and slows, the river drops what it carries. So anything that slows a river can make a new landscape. It could be a stick lodged against a stone, or the rib cage of a calf moose drowned at high water. Where the water piles against the obstacle, it drops its load, and an island begins to form. The island—in fact, any deposition—reshapes the current. And it doesn't have to be an obstacle; it could as well be a deflection. The river finds the course of least resistance. The energy reorganizes around that. Once the energy changes, everything changes.

And here's the point: No one pattern continues indefinitely; it always gives way to another. When there are so many obstacles and islands that a channel can no longer carry all its water and sediment, it crosses a stability threshold and the current carves a new direction. The change is usually sudden and often dramatic, the hydrologist said—a process called avulsion.

There, on the bench by the Alaskan river, morning light was beginning to seep between the mountains behind me, and the wet stones began to shine. It was going to be another hot day, but the morning was sweet. And there were the physics of the river, playing themselves out in front of me. A root-ball, dislodged from the bank, tumbled past me and jammed against a midriver stone. The current, dividing itself around the root-ball, wrinkled sideways and turned upstream. It curled into pocket eddies behind the roots. Even as I watched, the pockets filled with gravel and sand. A willow could grow there, and its roots could divide and slow the river further, gathering more gravel, creating a place where new life could take root. One obstruction had changed the river from an agent of destruction to a construction engineer, building the very structure that turned and slowed it.

I shoved a rock into the river. The sudden curl of current made me grin.

I don't have to stop the river. I don't have to lie awake at night in despair at the power of the economic and political forces rushing hell-bent toward runaway climate change. I don't have to ask my grandchildren to struggle to "adapt" to a world so much less than this one. My work and the work of every person who loves this world, *this one,* is to make one deflection in complacency, one obstruction to profits, one blockage to business-as-usual, then another, and another, to change the energy of the flood. As it swirls around these snags and subversions, the current will slow, lose power, eddy in new directions, and create new systems and structures that change its course forever. On these small islands, new ideas will grow, creating thickets of living things and lifeways we haven't yet imagined. Those disruptions can turn destructive energy into a new dynamic that finally reverses the forces that would wreck the world.

So we find our stone. And we chuck it in. Then we find another. This is the work of *conscientious refusal,* the steadfast refusal to let a hell-bent economy force us to row its boat. This is the work of *creative disruption,* turning the destructive energy of the fossil fuel industry into a new dynamic that replaces itself with something lasting. This is the work of *courageous, relentless citizenship,* pushing, requiring, demanding that elected government protect the commons: the atmosphere, the stable climate, the freshwater, the fertile soil, the great biodiversity of life, and this most important shared value, the future.

CONSCIENTIOUS REFUSAL: A STONE IN THE RIVER SLOWS THE FLOW

A stone is solid. There is no getting through it; the current has to slow down and go around. Many of us were alive when people said, "Hell, no," to an unjust war. Shoot, many of us *were* those people. The question is, can we say, "Hell, no," to an even more outrageously unjust industrial growth economy that is creating far more damaging consequences? If business-as-usual bullies us to live in ways we don't believe in, then we will stand up to the pressure. There are things we will refuse to do, understanding that there are costs to that refusal. This is a matter of claiming one's values as an act of resistance. Call it conscientious refusal. Call it *no*. Just no.

Every decision I make—how I travel, whether I travel, what I eat, what I invest in, whether I bear children, how many children I bear, what I teach them to love—embodies a moral value. My decisions speak for me. Yes, I believe in this. No, I do not believe in that and I will not participate. Making deliberate decisions at every point means that I cannot be suckered into carrying Big Oil's banner, suckered into being a foot soldier in Big Oil's war against the world.

No. If I am repelled by climate change, I can't eat beef cattle, those blowsy blowhards. If I'm horrified by the gyre of plastic in the middle of the Pacific, I can't buy My Little Pony. If I don't like the thought of Chinese children boiling out the heavy metals in a junk pile of discarded Dells, I have to get by on an old computer. If I am sickened by reports of oil slathered on the Arctic Ocean floor, I have to find a way to use dramatically less oil.

It's hard to break habits, and I'm nowhere near conscientious enough. Not even close. But like objectors in any other war, it outrages me to be made into an instrument of death and injustice. When all is said and done, I want to be able to say that I lived a life I believed in, not one that the lying oilmen and skillful marketers designed for me. Even if hope is rapidly failing that I can make a difference in the future of the Earth, I want to reclaim the power to make a difference in myself.

Is this sacrifice? Listen to Carl Safina, a marine ecologist:

> Dysfunctional values married to catastrophic leadership
> have led us to . . . believe that solution is sacrifice and that

sacrifice for a just cause is not noble but, rather, out of the question. . . . This refusal to "sacrifice" is actually a pathological refusal to change for the better. *That* is the real sacrifice. . . .

Nearly every just cause is a struggle between the good of the many and the greed of a few. But because greed has the advertising dollars to make selfishness fashionable, it sustains itself by turning enough people against our own self-interest. . . .

Of all the psychopathology in the climate issue, the most counterfunctional thought is that solving the problem will require sacrifice. As though our wastefulness of energy and money is not sacrifice. As though war built around oil is not sacrifice. As though losing polar bears, penguins, coral reefs, and thousands of other living companions is not sacrifice. As though withered cropland is not a sacrifice. . . . As though risking seawater inundation and the displacement of hundreds of millions of coastal people is not a sacrifice—and reckless risk. *But don't tell me we need a law mandating more efficient cars; that would be a sacrifice!* We think we don't want to sacrifice, but *sacrifice* is exactly what we're doing by perpetuating problems that only get worse; we're sacrificing our money, sacrificing what is big and permanent, to prolong what is small, temporary and harmful. We're sacrificing animals, peace, and children to retain wastefulness—while enriching those who disdain us.

If Carl is even half-right, then conscientious refusal can be a kind of liberation—a good joke on everybody's corporate handlers. Moreover, it can be *sacrifice* in the ancient sense: to *make sacred*, to give up something, some self-gratification, in order to bring oneself into closer relation to what is good in itself. Renouncing a hyper-consuming life, turning away from world-hurting practices, is a chance to think more clearly about what gives you true joy and moral courage.

Full disclosure: You will sacrifice out-of-season fruit shipped across the equator, winter vacations in Florida, spring skiing in Vail, irrigated crops and lawns, rapid travel, airport security lines, seat backs in an upright position, long commutes, poisonous potatoes, perfect apples, constant TV, isolation from neighbors, loneliness, guilt, and huge couches—and the psychic and carbon costs of these.

CREATIVE DISRUPTION: A STONE IN THE RIVER CHANGES THE FLOW

A stone is disturbing. It forces the river to do something different. We can say no, as in conscientious refusal, but we can also say yes. Creative disruptions are positive, creative acts that call attention to different courses forward. What sort of imaginative acts?

Technologies. Many of the disruptions in the river will be alternative ways of doing things that are better, cheaper, safer, and more successful than fossil-fuel-driven technologies. Heat your home without burning anything up, and you've weakened the fossil fuel economy. Transport yourself without burning anything up, and you've introduced a perturbation into the system. Feed yourself without burning anything up, and suddenly it isn't so profitable to blow up a seam of coal. Generate electricity on your roof, and you've reduced the hegemony of the centralized, coal-fired electricity grid.

Innovations of every sort can change the economics of fossil fuels. The assumption is that a good idea drives out a bad one. If there is a better way, the river will flow toward these and away from coal and oil, the industry of the smudged past. Innovations can be very good rocks. This is a call for people of systematic imagination to *find a better way.*

Arts. The creative imagination in play has a profoundly disruptive effect on the old ways, because it exposes the lies and calls attention to truths we might have forgotten. Music, theater, children's choirs, literature—all the beautiful human expressions of grief and decency and celebration are ways to tell the truth, for god's sake, about the moral consequences of an oil-fueled future. We have to face the facts, in every way we know how. This is a call to the arts—to move from the galleries into the streets and wash out the lies.

Street theater. If the truth is that Big Coal's coal trains will destroy the dreams of children, then paper the rails with photos of children, so the engineer has to make a conscious decision to grind over their little faces. If the truth is that coal dust will make people sick, send the black-robed choir to sing the requiem at the railroad crossing. This is the hearse dogging the fracking trucks. These are the roadside funeral bouquets at the turnoff to the logging site. These are the billboard postings of family photos from the senator's home page: Would you sell out *these* children? Or how about this: billboard photos of congressional "representatives" wearing the corporate logos of all their sponsors? The bigger the check, the bigger the logo—as Bill McKibben suggested.

Street art and theater make abstract truths *manifest*—palpable, touchable, from *manus,* the hand. This is the goal. Manifestation is particularly important, because the emergencies we face are invisible by their nature: plants that are not there, animals that are not there, undernourished babies not yet born, islands sunk beneath rising seas, silences, *the silences*, wars not yet begun, melted glaciers, carbon dioxide itself—odorless, colorless, a fearsome ghost that swirls its skirts, unseen, around the planet.

Investigative journalism. Let's insist on knowing truths that have been hidden. What are the routes of the coal trains? What are the chemicals in fracking injections? What are the health risks of atrazine and every other estrogenic chemical sprayed on our food? What has happened to the oil that gushed into the Gulf of Mexico, Prince William Sound, Puget Sound, the Kalamazoo River, and all the unnamed others? And why, and by what human agency, are these facts kept from us? The truth will set you free? Perhaps—and if we're lucky, it will put some corporate CEOs in jail.

Direct action. Creative disturbance is a call to *get in the way.* To get in the way of the coal trains. To get in the way of the pipelines. To get in the way of the corporate meeting. To make it harder for corporations to do the wrong thing. So, as it must in the face of any massive moral failure of government, creative disruption may require direct action and civil disobedience. So this is a call to imagine courage. Courage is not fearlessness. "I'm afraid all the time," biologist Sandra Steingraber said as she was led to jail yet again for protesting fracking. "But I do it anyway." That's courage: to be afraid, and to take action all the same. Climate courage: Let's call it that, and let's notice that Sandra's courage, and that of her allies, stopped fracking in the state of New York.

My students also teach me about courage. Some time ago, the phone rang in my university office. It was a student who had graduated the year before, a direct-action activist whom I expected to read about in the papers. "It's about a broken heart," he said. "You go to the forest and you see trees that have been growing for two hundred years. Cut to stumps. You look at this, and you say, *This just has to stop*."

He was talking fast, calling me on his cell phone from an unnamed place. I could hear him trying to catch his breath. "The only thing you can do is put your body between the trees you love and the machines that destroy them. You put it all on the line—your body, your livelihood, your reputation. Maybe your life."

The forest he loved was a spirit-lifting place of slanting light and slippery trails, all smelling of sweet bracken fern and pine. The last time I was there, I waded through wildflowers between rough trunks so tall that they disappeared into fog. It was a sobering thing when much of that forest burned, but more terrible still when chainsaws felled the trees that the fire had left standing, the living and the dead. When bulldozers shoved the charred spars into burn piles, they took away even the nourishment that might let a forest flourish in that place again. When loggers cut the living trees, they took away the hot spots of life that would spread back over the burnt land. What could be more defenseless than a fire-scarred forest? What could be more sublime? What is a person supposed to do when the machines move in to reduce ancient trees to profit, ash, dust, and global warming?

Just the year before, that same young man was part of my environmental ethics course. Together, we debated the moral justification for civil disobedience and direct action. We read and reread Saint Augustine. "An unjust law is no law at all," he wrote, and we struggled to understand our obligations. But the world is complicated. It's hard to know what's right and what's wrong, and too great a moral certainty is its own kind of injustice. Cedars, huckleberries, birdsong—entire forests fell as we reframed the questions, reducing us to a silence close to despair. Only gradually did I begin to understand the moral power of that sorrow. Maybe civil disobedience isn't about justice and obligation. Maybe it's about love. Maybe, my student suggested, we're called to act in defense of the green growing world not from an obligation to disobey laws that are destructive but from love for what is being destroyed.

Sandra Steingraber wrote: "We are all members of a great human orchestra, and it is now time to play the Save the World Symphony. You do not have to play a solo, but you do have to know what instrument you hold and find your place in the score." The metaphor is beautiful and strong, just like Sandra. But the really exciting, the truly hair-raising, part of all this is that there is no score. No one has written the Save the World Symphony. We're making it up as we go along. This is not classical music. This is jazz, with all its risk and glory.

COURAGEOUS, RELENTLESS CITIZENSHIP: A STONE IN A RIVER IS ONE OF MANY

A single stone is just a heavy gray thing. A course of stones is a riverbed. In the hydrology of activism, collective action is essential—and so we come to the work of citizenship. The purpose of government, in my view, is to be the mechanism by which people work together to do things that none of them can do alone. In my state, Oregon, for example, the first settler government began as a meeting of neighbors to figure out how to work together to keep the wolves from the sheep. The problems of climate change and ecological collapse seem to me to be quintessential examples of the necessity of government. Individual people (including corporations) pursuing their perceived self-interests with few restraints have created a disaster. None of us can make a dent in the crisis by ourselves. So we turn to government, charging it with the responsibility to protect the ecological and geophysical commons on which the exercise of individuals' basic rights depends. The failure of government to act effectively on climate change is cataclysmic, as it marks the failure of a people with shared critical interests to bring themselves into concerted and coordinated action.

So what do you do when the reality is so far from the ideal?

You call your government to account. A single voice is, of course, a thin thread. A chorus of voices, however, is a movement. Here's what I say: Put your elected representatives on speed dial—hometown, state capital, Washington, DC. Give those sullen, frightened folks some spine. They know what's right. They know what's wrong. Business-as-usual is suddenly immoral. Dumb is unforgiveable. Delay is unconscionable. Selling out your constituents is monstrous. If Big Oil has soaked congressional representatives in natural gas and oil dollars, then it's up to us to hold their feet to the fire.

In a gerrymandered, corporate-sponsored, dark-moneyed plutocracy, I'm not sure how much meaning is left in elections. Nonetheless, I vote faithfully and campaign for candidates I trust. But there's an additional way to take a stand, and that is to, well, *take a stand.* Again and again, history changed course when people in large numbers took to the street and stood up to be counted. Maybe that's a better way to vote.

MAY I ESPECIALLY address the grandparents now? You are, after all, my compatriots. It's time for a frank talk. This is given: We love our children and grandchildren more than life itself. We would, in literal fact, do *anything* for them.

This also is given: Grandparents are in a powerful position to protect the children and grandchildren. The first asset we bring to the work is a set of skills, experience, and knowledge gained over a lifetime of productive work. The second asset is political clout. We vote, and there are a lot of us. We donate to political campaigns, and we have a lot to give. The third asset is something that we have in abundance that no other demographic has enough of: We have time. Put these assets together and grandparents command the power to shape the new world. We can make sure that our children and grandchildren inherit a planet that will sustain their health and nourish their freedom to make a good life.

How? By organizing our huge political power to elect officials who will get down to the most important business of protecting the life-sustaining systems of the planet. Knock on doors for these politicians, then hold them to account. If politicians dither or jabber or make excuses, send them home. Organize for truly fast and effective climate change legislation. Organize for solar power. Organize to keep poisons from the air, the water, the agricultural fields. Tell AARP it's time to stop worrying about the health of our colons and golf games and to start worrying instead about our legacy. Focus the church education program on environmental toxins instead of the nature of heaven. Get the alumni association off the cruise ships and into the streets with banners.

Get started. At this stage in our lives, it mocks death to waste time. We work all our lives to provide for the future of our children and grandchildren. We cannot let it all slip away in our last decades. A life-sustaining planet, not an MP3 player or a plate of cookies, will be our last and greatest gift to the ones we love the most. Without that gift, all other gifts are meaningless, and the hugs of a grandparent become cynical jokes on the beautiful little ones, who do not deserve the struggles they will face.

PS: Let us hear no more talk about the extra entitlements of the elders—entitlements to year-round perfect weather, an annual trip to Las Vegas, low taxes, easy Sunday crosswords, reduced greens fees and the world be damned. It may be true that because we have worked hard, we deserve to enjoy the fruits of our labor; but we owe those fruits also to a stable climate, temperate weather, abundant food, cheap fuel, and a sturdy government—all unearned advantages that our children will not have if we don't act. It's tragic and culturally dysfunctional if our lives culminate in a radical selfishness that makes us angry and bitter (because selfish desires can never be satisfied) rather than in the respect that we truly earn as people of wisdom and responsibility.

YOU AND I and the young red-haired woman in the auditorium in San Jose are not the first to ask what one person might do, and—god help us—we will not be the last. We understand that we are part of a global movement that will never end but that will struggle generation after generation under increasingly desperate conditions, until humanity either finds a way to live justly and joyously on Earth or fails. In that case, it will be other species that will carry on.

Toward the end of her life, the courageous advocate of women's suffrage, Elizabeth Cady Stanton, wrote to her comrade-in-arms Susan B. Anthony: "This is winter wheat we're sowing, and other hands will harvest." That inspired the beautiful songwriter Libby Roderick to write "Winter Wheat":

When I was young I thought that failure was impossible
All wrongs would be righted in my time.
Now I am old I see that failure is impossible
I pass the torch to you. Will you hold it high?

For we are sowing winter wheat
That other hands will harvest
That they may have enough to eat
After we are gone.

We will plant shade trees that we will not sit under
We will light candles that others can see their way
We'll struggle for justice though we'll never see it flower
Our children's children will live in peace one day. . . .

Each generation passes like the leaves
On an old oak tree whose roots are strong.
Each new generation bursts out like the spring
And they will be the ones to carry on.

When I was young, I dreamed the Earth was healed and
whole again,

Creatures, trees, and rivers free and wild
Now I am old I dream the planet healed and whole again
That dream's reborn forever in the heart of each new child.

So we are sowing winter wheat
That other hands will harvest
That they might have enough to eat
After we are gone.

We will plant shade trees that we will not sit under
We will light candles that others can see their way
We'll struggle for justice though we'll never see it flower
Our children's children will live in peace one day.

after hope, the roar of the lion, the great rising wave

AND SO WE come to the obligatory question about hope. "Is there a reason to hope?" asked the bearded man who had waved his hand like a flag, and the white-haired woman with her grandson in tow, and the college student in a football jersey. The question seems to be on the mind of every person in every audience.

"What do you mean by the question?" I asked. That's a response that makes people hate philosophers.

Do you mean, "Is there good reason to think that the Earth can avoid sweeping extinctions and disruption of the systems that support our lifeways?" If that's the question, the answer increasingly is "No, not really." This is because significant planetary warming is locked in by the greenhouse gases already in the atmosphere. Right now, the exquisitely sensitive planet has warmed 0.8 degrees Celsius, and already glaciers and ice sheets are melting, weather patterns are shifting, and droughts, flooding, and heat waves are increasing in number and intensity. Even if the world stops emitting greenhouse gases today, the temperature of the Earth will rise 1.5 degrees Celsius above preindustrial levels, almost double the increase we have already seen.

Gus Speth, the former dean of the Yale School of Forestry, wrote, "All we have to do to destroy the planet's climate and ecosystem and leave a ruined world to our children and grandchildren is to keep doing exactly what we are doing today." Bad enough, but he's not quite right. From the new figures, it seems that all we have to do to assure this awful outcome is exactly nothing. What we have already done is apparently adequate to the task. On the other hand, if we *do* keep doing exactly what we are doing today, new reports from the IPCC say, there is no preventing the temperature from rising by 3.6 degrees Fahrenheit in the next three decades. The *New York Times*: "According to a large body of scientific research, that is the tipping point at which the world will be *locked into* [emphasis mine] a near-term future of drought, food and water shortages, melting ice sheets, shrinking glaciers, rising sea levels and widespread flooding."

Climate activists have gotten the message. New polling from the Yale Project on Climate Change Communication finds that, of those who are "alarmed" or "concerned" about climate change, 85 percent feel afraid, 81 percent feel sad, 79 percent are angry, 76 percent are disgusted, and—worst of all—61 percent feel helpless. I don't hear much hope in those numbers. As the nineteenth-century German philosopher Arthur Schopenhauer wrote, "Hope is a wish that we doubt will come true."

But hope isn't just about reasons, is it? It's more complicated than that. Hope may be an emotion that runs counter to reason. If we had *reason* to hope, we wouldn't need hope, right? We could just go with the evidence. Hope is something more, something far more important and intriguing. If the question is, "Should I feel hope?" my answer can only be that Western philosophy has been struggling with the question ever since Pandora opened her jar and released evils into the world.

Pandora knew she was forbidden to open the jar, but all she wanted was a little peek. She raised the lid. Death, Despair, Greed, War, Madness, Pestilence, and Betrayal flew out shrieking like crows from a swampy night roost, pulled at Pandora's hair, and sent her diving into the bushes. I can imagine that Pandora slammed the lid back on and looked around wildly, her heart racing. When she calmed down, I imagine that she shook the jar to make sure it was empty. It was not. Stuck in the bottom was Hope.

Here, accounts differ. Some say she never opened the jar again—surely she wouldn't—and hope remains forever denied to mortals. Others say she opened the jar—reckless mortal. Out fluttered winged Hope. Hope: how lightly it must have flown on its feathered wings, how delicately it landed on Pandora's outstretched hand. She didn't understand its horror or its blessing.

And here, again, accounts differ. In one telling, Pandora released the final evil, the one thing that empowers all the evils of the world. "In reality [hope] is the worst of all evils," Friedrich Nietzsche wrote, "because it prolongs the torments of man." If we had no hope, nothing could harm us. We could choose to die by our own hands and by that act, frustrate all our demons.

Accordingly, French existentialist Albert Camus compared hope to the painted screen that executioners once held in front of the faces of prisoners to hide their view of the scaffold as they climbed the stairs. Instead of black crows hunched on the gallows pole, prisoners saw lively swallows darting over Italian hills arrayed with vineyards and poplar trees. Hope is in this account a delusion, the vision of something better that keeps us climbing toward our inevitable deaths.

The ecophilosopher Joanna Macy acknowledges this philosophy of despair and goes beyond it. Hope is exactly what Camus described: a vision of something better that keeps us climbing. But it's not a delusion. It's a radical imagining, a courageous affirmation of what a person values too much to let die. The vision might not be present to his eyes, actualized in the landscape, but it is vividly alive in his ability to imagine something different. Macy calls this vision active hope. In this telling, hope empowers all the good in the world, keeping us mortals climbing the stairs toward a vision of the better, which is what we do so nobly and what has become our art, our beauty, our cause for celebration.

In her book, also called *Active Hope,* Joanna frames hope as a kind of process thinking, a movie that changes from frame to frame to create change over time. "If something is not in the picture at the moment, that doesn't mean it won't be later on. This way of conceiving reality sees existence as an evolving story rather than as predefined. Because we can never know for sure how the future will turn out, it makes more sense to focus on what we'd like to have happen, and then to do our bit to make it more likely. That's what Active Hope is all about." She describes the film like this:

> [Frame one] This is how things are now
>
> [Frame two] But choices I make influence what happens next
>
> [Frame three] So what is my hope [I think she means, vision of the good]? And how can I be active in moving toward that?

Continuing the film analogy: To engage in active hope, people have to make space in their lives, quite intentionally, to envision an inspiring future—what Macy calls the "desired destination"—to map out the story of how to get there, and then to identify their role in the story. Active hope is hugely important in our time—not despite the fact that but *precisely because* the planet is entering into a time of disruption and dying. It is active hope that can make this time the best it can be rather than the worst.

So yes, seek out active hope. Or, as Barbara Kingsolver said, "If you run out of hope at the end of the day . . . rise in the morning and put it on again with your shoes." If it sustains us in the work of making the best of things—literally *making the best of things* from a realistic assessment of the options—hope may be the most important virtue of our time. Virtues are character traits. Character traits are habits of the mind. So it is a very good thing to get in the habit of hope.

However, it turns out that the bearded man in the audience had quite a different interest in hope. When he clarified his question, this is what he said: "There is no hope. Nothing I do will make any difference. I can't save the world from climate change or ecological collapse. So I'll just keep on buyin' and burnin' the way I always have. There's no point in change to no effect." What do I say to that?

Well, like any philosophy professor, I called him out: "What the heck kind of reasoning is that? You don't do the right thing because it will have good results. You do the right thing because you believe it's the right thing." What would you say to a slave owner who made that kind of argument? *Alas, I could free every one of my slaves and it wouldn't make a dent in the slave trade. The institution of slavery is so much bigger than one little owner. So I'll just keep on working these people in the day and chaining them up at night. No point in change to no effect.* What would you say? You'd say, "It doesn't matter if you can or cannot change the world. What matters is that you can change yourself."

It is arguments like this that give hope a bad name. Derrick Jensen and Michael P. Nelson have made the argument that at this point in time, the world might be better off without hope. Let's follow the reasoning.

Hope is a particularly Western notion. It's no accident that Hope emerged from an ancient Greek jar. Hope would probably not have flown from a Japanese ginger jar, because a central teaching of many Eastern traditions, Zen Buddhism, for example, is to live in the moment, not in hope or fear of the next day. But most of us in the U S of A? We're so invested in what will happen next that we hardly notice the light in the trees or the sun's warmth on our shoulders. This focus has created a bizarre moral tradition. It's an aberration in the moral history of the universe. But because it has infused our ways of thinking, we think it's the normal—or the only—way to think. The name of the tradition is consequentialism (here it is again; we have seen it mucking things up before), and it measures the moral worth of an act by a calculation of its consequences in the future.

If we calculate right and wrong by this complicated cost–benefit analysis, then we have to be always "fixated on the future," as Nelson writes, "perpetually . . . justifying means by their ends. . . . We have therefore built a society that can be readily disempowered"—not by hopelessness but by this bizarre idea that the only acts worth doing are those that will have some sort of payoff.

Consider the moral abdication of blind hope: If everything is going to turn out okay, regardless of what I do, then I am under no obligation to do anything, because nothing I do will make a difference. Consider the moral abdication of blinding despair: If everything is going to go to hell, regardless of what I do, then I am under no obligation to do anything, because nothing I do will make a difference. Blind hope or blinding despair? Either way, I don't have to do anything. I'm off the hook.

But notice that this is a logical fallacy—the fallacy of false dichotomy. It's not true that our only options are hope and despair. Between them is a broad middle ground, which is acting from neither hope nor despair; it has nothing to do with future consequences. This broad middle ground is acting with *integrity*. Integrity: like integer, a whole number; from *integritatem*, wholeness, a consistency between what one believes and what one does. Poet Robinson Jeffers wrote:

> . . . Integrity is wholeness, the greatest beauty is
>
> Organic wholeness, the wholeness of life and things, the
> divine beauty of the universe. Love that, not man
>
> Apart from that, or else you will share man's pitiful confusions, or drown in despair when his days darken.

The times call for integrity. The times call for the courage to refute our own bad arguments and renounce our own bad faith. We are called to live lives we believe in, even if a life of integrity is very different—let us suppose radically different—from how we live now. People of integrity live gratefully because they believe that life is a gift. They act reverently because they believe the world is sacred. They live simply because they don't believe in taking more than their fair share. They act lovingly toward the world because they love it. As a person of integrity you'll find your work at "the place where your deep gladness and the world's deep hunger meet," as theologian Frederick Buechner wrote.

Standing at the podium, trying to steady my voice, trying to keep from crying, here's what I said to the bearded man who asked about hope: Each of us has the power to make our life into a work of art that expresses our deepest values. Don't ask, *Will my acts save the world?* Maybe they won't. But ask, *Are my actions consistent with what I most deeply believe is right and good?* This is our calling—the calling for you and me and everybody else in the room: to do what is right, even if it does no good; to celebrate and care for the world, even if its fate breaks our hearts.

And here's the paradox of hope: that as we move beyond empty optimism and choose to live the lives we believe in, hope becomes transformed into something else entirely. It becomes stubborn, defiant courage. It becomes principled clarity. And when courageous-hearted, clear-minded people find one another, it becomes a powerful creative force for social change. We've seen this again and again: slaves leading slaves toward the North Star, crowds singing as they march across a bridge, mothers baring their breasts at the gates of the prisons that hold their sons, young people refusing to go to war—great rising tides of affirmation of justice and human decency and shared thriving.

I HAVE SEEN tides build on the beach in a wild cove. The water creeps in slowly at first. Almost imperceptibly, the rising tide encircles small stones, lifts rock wrack, bends green eelgrass toward shore. An unexpected swell porpoises over the back of the incoming tide and booms against a rocky ledge. A geyser and smatter of water. Then there are lines of breakers advancing through the inlet, one after another, filling the tide that flows in full force now. As the tide charges toward small ducks, harlequins and scoters, they startle and dive under the wave. The tide advances toward the river that pours down toward the sea. There is power in the river, but the tide meets it face-on. The river and the tide butt against each other; there are dangerous currents here, and choppy waves. But slowly, and then with gathering speed, the single roaring wave of the tide pushes the river back, and back again. Behind the tidal bore, the cove is full and shining.

AFTERWORD

I DON'T KNOW WHAT the future will bring. Standing here in the middle of seven generations, I can look back three generations to imagine what life was like for my great-grandparents. Even though they lived through world-shattering wars, the fundamental presuppositions of life survived. Spring would follow winter, and the rains would return, vultures would come to the battlefields, and the poppies would bloom. Now that the world has entered a time of discontinuities and abrupt, cascading change, it's harder for me to imagine the lives of my great-grandchildren, three generations into the future. The best I can do is try to imagine standing on my Oregon land in 2025, ten years from today, thinking over the events of those intervening years.

RING THE ANGELUS

Dateline: May 25, 2025, Wren, Oregon. All those years, the Swainson's thrushes were the first to call in the mornings. Their songs spiraled like mist from the swale to the pink sky. That's when I would take a cup of tea and walk into the meadow. Swallows sat on the highest perches, whispering as they waited for light to stream onto the pond. Then they sailed through the midges, scattering motes of wing-light. Chipping sparrows buzzed like sewing machines as soon as the sun lit the Douglas firs. If I kissed the knuckle of my thumb, they came closer and trilled again.

For years there were flocks of goldfinches. After my husband and I poisoned the bull thistles on the far side of the pond, the goldfinches perched in the willows. When they landed there, dew shook from the branches into the pond, throwing light into new leaves where chickadees chirped. The garbage truck backed down the lane, beeping its backup call, making the frogs sing, even in the day.

Oh, there was music in the mornings, all those years. In the overture to the day, each bird added its call, until the morning was an ecstasy of music that faded only when the diesel pumps kicked on to pull water from the stream to the neighbor's bing cherry trees.

Evenings were glorious too. Just as the sun set, little brown bats began to fly. If a bat swooped close, I heard its tiny sonar chirps, just at the highest reach of my hearing. Each downward flitter of its wings squeezed its lungs and pumped out another chirp, the way a pump organ exhales Bach. Frogs sang and sang, but not like bats or birds. Like violins, violin strings just touched by the bow, the bow touching and withdrawing. They sang all evening, thousands of violins, and into the night. They sang while crows flew into the oaks and settled their wings, while garter snakes, their stomachs extended with frogs, crawled finally under the fallen bark of the oaks and stretched their lengths against cold ground.

I don't know how many frogs there were in the pond then. Thousands. Tens of thousands. Clumps of eggs like eyeballs in aspic. Neighborhood children poked them with sticks to watch their jelly shake. When the eggs hatched, there were tadpoles. I have seen the shallow edge of the pond black with wiggling tadpoles. There were that many, each with a song growing inside it and tiny black legs poking out behind. Just at dusk, a hooded merganser would sweep over the water, or a pair of geese, silencing the frogs. Then it was the violins again, and geese muttering.

In the years when the frog choruses began to fade, scientists said it was a fungus, or maybe bullfrogs were eating the tadpoles. No one knew what to do about the fungus, but people tried to stop the bullfrogs. Standing on the dike, my neighbor shot frogs with a pellet gun, embedding silver BBs in their heads, a dozen holes, until she said, *How many holes can I make in a frog's face before it dies? Give me something more powerful.* So she took a shotgun and filled the bullfrogs with buckshot until, legs snapped, faces caved in, they slowly sank away. Ravens belled from the top of the oak.

When the bats stopped coming, they said that was a fungus too. When the goldfinches came in pairs, not flocks, we told each other the flocks must be feeding in a neighbor's field. No one could guess where the thrushes had gone.

Two springs later, there were drifts of tiny white skins scattered in the shallows like dust-rags in the dusk. I scooped one up with a stick. It was a frog skin, a perfect empty sack, white, intact, but with no frog inside—cleaned, I supposed, by snails or winter—and not just one. Empty frogs scattered on the muddy bottom of the pond. They were as empty as the perfect emptiness of a bell, the perfectly shaped absence ringing the Angelus, the evening song, the call for forgiveness at the end of the day.

As it happened, that was the spring when our granddaughter was born. I brought her to the pond so she could feel the comfort

I had known there for so many years. Killdeer waddled in the mud by the shore, but even then, not so many as before. By then, the pond had sunk into its warm, weedy places, leaving an expanse of cracked earth. Ahead of the coming heat, butterflies fed in the mud between the cracks, unrolling their tongues to touch salty soil.

I held my granddaughter in my arms and sang to her then, an old lullaby that made her soften like wax in a flame, molding her little body to my bones. *Hush-a-bye, don't you cry. Go to sleep you little baby. Birds and the butterflies, fly through the land.* I held her close, weighing the chances of the birds and the butterflies. She fell asleep in my arms, unafraid.

I will tell you, I was so afraid.

Poets warned us, writing of "the heart-breaking beauty [that] will remain when there is no heart to break for it." But what if it is worse than that? What if it's the heartbroken children who remain in a world without beauty? How will they find solace in a world without wild music? How will they thrive without green hills edged with oaks? How will they forgive us for letting frog song slip away? When my granddaughter looks back at me, I will be on my knees, begging her to say I did all I could.

I didn't do all I could have done.

It isn't enough to love a child and wish her well. It isn't enough to open my heart to a bird-graced morning. Can I claim to love a morning if I don't protect what creates its beauty? Can I claim to love a child if I don't use all the power of my beating heart to preserve a world that nourishes children's joy? Loving is not a kind of *la-de-da*. Loving is a sacred trust. To love is to affirm the absolute worth of what you love and to pledge your life to its thriving—to protect it fiercely and faithfully, for all time.

My husband and I were there when the last salmon died in the stream. When we came upon her in the creek, her flank was torn and moldy. She had already poured the rich, red life from her muscles into her hopeless eggs. She floated downstream with the current, twitching when I pushed her with a stick to turn her upstream again. Sometimes her jaws gaped, still trying to move water over her gills. Sometimes she tried to swim. But she bumped against rocks, spilling eggs onto the stones. Without reason, she pushed her head into the air and gasped. We waded beside her until she died. When she was dead, she floated with her tail just above the surface, washing downstream until she lodged on a gravel bar. The music she made was the riffle of rib bones raking water, then no sound at all as her body settled to the bottom of the pool.

I buried my face in my hands, even as I stood in the water with the current shining against my shins. Oh, we had known the music of salmon moving upstream. When the streams were full of salmon, crows called again and again, and seagulls coughed on the gravel bars. Orioles sang, their heads thrown back with singing. Eagles clattered. Wading upstream, we walked through waves of carrion flies, which lifted off the carcasses to swarm in our faces, buzzing like electrical current. Water lifted and splashed, swept by strong gray tails, and pebbles rolled downstream. It was a crashing coda, the slam and the buzz and the gull-scream.

RING THE ANGELUS for the salmon and the swallows. Ring the bells for frogs floating in bent reeds. Ring the bells for all of us who did not save the songs. Holy Mary, mother of God, ring the bells for every sacred emptiness. Let them echo in the silence at the end of the day. Forgiveness is too much to ask. I would pray for only this: that our granddaughter would hear again the little lick of music, that grace note toward the end of a meadowlark's song.

Meadowlarks. There were meadowlarks. They sang like angels in the morning.

NOTES

PREFACE: AT LOW TIDE, WATCHING THE WORLD GO AWAY

This piece was published in different form in *High Country News* (September 1, 2014), 26.

The warning from Stanford scientists is in Anthony D. Barnosky and more than five hundred others, "Scientific Consensus on Maintaining Humanity's Life Support Systems in the 21st Century: Information for Policy Makers," 2013, http://mahb.stanford.edu/consensus-statement-from-global-scientists/.

Charles Taylor writes in his book *The Ethics of Authenticity*, 22.

My work about climate change as a moral issue began as a rich and continuing collaboration with my colleague Michael P. Nelson, the Ruth H. Spaniol Chair of Renewable Resources at Oregon State University. A number of the ideas in this book were generated in conversations with Michael and draw on our coedited book, *Moral Ground: Ethical Action for a Planet in Peril* (San Antonio: Trinity University Press, 2011).

WHY IT'S WRONG TO WRECK THE WORLD

The convening of environmental philosophers was one of an annual series of Blue River Writers events sponsored by the Spring Creek Project for Ideas, Nature, and the Written Word at Oregon State University (Charles Goodrich, director).

Aristotle explains the practical syllogism in *Nicomachean Ethics*.

The response of professional philosophers to our climate change use of the practical syllogism has been interesting. They want to be sure we know that the logic is more complex than what we have presented, and of course it is. There are values deeply embedded in the first premise, the statement of fact; and there are facts deeply embedded in the second premise, the statement of values. We are grateful to them for pointing this out in print. I am grateful to the discipline of philosophy for the care and precision of its language and work, driving ever deeper into hard and complex questions. This work is essential and important, providing clarity where muddleheadedness has been disastrous, providing new ways to think about problems whose old solutions now fail us, showing a courage to take on moral questions so difficult that even politicians recoil from them. This is good work, done by good and exceptionally smart people. And I should say that digging deep, deep into a question is a great intellectual pleasure besides—a pleasure I have shared with many philosopher colleagues. That said, I have chosen to write a book that is different from the book my colleagues would have written—different in audience and purpose—and I hope they will allow me that departure from academic business-as-usual.

Michael P. Nelson and I cowrote "Thirteen Good Reasons to Save the World," drawing on the introduction to *Moral Ground.*

BECAUSE THE WORLD IS WONDERFUL

Readers will love Abraham Joshua Heschel's book *Man Is Not Alone: A Philosophy of Religion* (New York: Farrar, Straus and Giroux, 1976). The quotation about exaltation is on page 15.

Mary Evelyn Tucker's ideas are beautifully captured in *Journey of the Universe,* the film and companion book (New Haven: Yale University Press, 2011) she made with her collaborators Brian Swimme and John Grim. The quotes come from my interview with Tucker, "A Roaring Force from One Unknowable Moment," *Orion* 34, no. 3 (2015).

Readers can find the reference to the world's lyricism in Thomas Berry, *The Great Work: Our Way into the Future* (New York: Broadway Books, 2000), 49—and then they'll want to read the entire book.

Pinning down statistics about rates of extinction is like trying to nail down the tail of a hurricane—they are changing every moment. I've drawn these from the World Wildlife Fund, as reported by Damian Carrington, "Earth Has Lost Half Its Wildlife in the Past 40 Years, Says

WWF," *Guardian*, September 30, 2014, http://www.theguardian.com/environment/2014/sep/29/earth-lost-50-wildlife-in-40-years-wwf.

The sociologist Mary Catherine Bateson is a member of the Mellon Foundation's Council on the Uncertain Human Future, a gathering of twelve women, of whom I am one. She made these remarks in a meeting on May 8, 2014.

BECAUSE WE LOVE THE CHILDREN

Brian Doyle is a Portland poet, essayist, editor, and father. He handed me these words on a folded piece of paper after his talk at the Magic Barrel readings in Corvallis, Oregon, in November 2013. I reprint them here with his permission.

BECAUSE WE HONOR HUMAN RIGHTS AND JUSTICE

Sheila Watt-Cloutier's words are adapted from the transcripts of Indigenous Peoples' Resistance to Economic Globalization: A Celebration of Victories, Rights and Cultures (New York, November 23, 2006).

The quote from Nicholas Stern appears in a *Guardian* article by Robin McKie, "Climate Change 'Will Make Hundreds of Millions Homeless,'" May 11, 2013, http://www.theguardian.com/environment/2013/may/12/climate-change-expert-stern-displacement.

For a full account of John Rawls's understanding of justice, see his work *A Theory of Justice* (New York: Belknap, 1999).

I first read Peter Singer's account of climate justice in Peter Singer, "A Fair Deal on Climate Change," *Project Syndicate*, June 10, 2007, http://www.project-syndicate.org/commentary/singer24.

Derrick Jensen's analogy of the aliens can be found in full in the piece "You Choose," in *Moral Ground*, 63.

A LOVE STORY

Anne Lamott's *Bird by Bird: Some Instructions on Writing and Life* (New York: Anchor Books, 1995) is lovely and useful. That's where I found her father's advice to her brother and to us all.

A story is not "a thing, after all, but an endless series of single acts." The quoted passage comes from a character in Richard Ford's novel *Let Me Be Frank with You* (2014), reviewed in the *New York Times Book Review* by Jonathan Miles (November 13, 2014).

I wrote "At the East Fork Cabin, All Is Well" as part of the Artist-in-Residence Program of the National Park Service.

The list of what it means to love a place is adapted from my essay "What It Means to Love a Place," in my book *The Pine Island Paradox: Making Connections in a Disconnected World* (Minneapolis: Milkweed, 2004), 34.

ON JOYOUS ATTENTION

My essay about bears is adapted from "Bear Sign (On Joyous Attention)," first published in *The Way of Natural History*, ed. Thomas Lowe Fleischner (San Antonio, TX: Trinity University Press, 2013).

The formal argument for the import of joyous attention is adapted from *Moral Ground*, 329–330, and shared here with permission of Trinity University Press.

Rachel Carson's important book is *The Sense of Wonder* (New York: Harper, 1998). Read it; give copies to all the parents in your lives.

Carson writes about the significance of the little ghost crab in *The Edge of the Sea* (New York: New American Library, 1955), 5–7.

Find Philip Cafaro's argument in his essay "Thoreau, Leopold, and Carson: Toward an Environmental Virtue Ethics," in *Environmental Ethics*, ed. Ronald Sandler and Philip Cafaro (Oxford: Rowman and Littlefield, 2005), 31–44.

The ideas about Rachel Carson draw on my essay "The Truth of the Barnacles: Rachel Carson and the Moral Significance of Wonder," in the book I edited with Lisa H. Sideris, *Rachel Carson: Legacy and Challenge* (Albany: State University of New York Press, 2008).

I wrote "The Art of Watching" in the Artist-in-Residence Program of the National Park Service. Find it at Denali National Park—you can get a copy from the bus drivers, who hand it out to their passengers.

AN OLD WORLDVIEW, A NEW WORLDVIEW

Thomas Berry's words come from *The Great Work,* 1–3. But I direct readers to a brand-new book that provides a broad look at Berry's inspiring work. It's *Thomas Berry: Selected Writings on the Earth Community*, ed. Mary Evelyn Tucker and John Grim (Maryknoll, NY: Orbis, 2014).

For more about Joanna Macy and the Great Turning, I recommend you start with Joanna Macy and Molly Young Brown, *Coming Back to Life: The Updated Guide to the Work That Reconnects* (Gabriola Island, BC: New Society Publishers, 2014). The words I quoted are on Joanna's website: www.joannamacy.net.

Readers will be well rewarded by exploring the website of the singer and songwriter Libby Roderick: www.libbyroderick.com. To date, she hasn't yet created a recording-quality version of this song; maybe you'd like to send her money so she can. I have reprinted the lyrics with her permission.

The material about the varieties of kinship is adapted from my essay "Late at Night, Listening," in *The Pine Island Paradox*, 50.

AN ETHIC OF THE EARTH

Henry Beston's *The Outermost House: A Year of Life on the Great Beach of Cape Cod* (New York: Holt, 2003) is a lovely book, wise on the gifts of the Earth, as in the short excerpt.

The Blue River Declaration is the work of the Blue River Quorum, convened by the Spring Creek Project for Ideas, Nature, and the Written Word in the ancient groves of the HJ Andrews Research Forest in the Blue River watershed in Oregon. The participants in that epic writing project were philosopher J. Baird Callicott, development officer Madeline Cantwell, poet Alison Hawthorne Deming, feminist philosopher Kristie Dotson, poet Charles Goodrich, ecopsychologist Patricia Hasbach, novelist Jennifer Michael Hecht, marine ecologist Mark Hixon, botanist and writer Robin Wall Kimmerer, philosopher Katie McShane, forest ecologist Nalini Nadkarni, philosopher Michael P. Nelson, student Harmony Paulsen, sociologist Devon G. Pena, singer/songwriter Libby Roderick, novelist Kim Stanley Robinson, geologist Fred Swanson, religion historian Bron Taylor, philosopher Allen Thompson, philosopher Kyle Powys Whyte, ecocritic Priscilla Solis Ybarra, religion philosopher Gretel Van Wieren, and writer Jan Zwicky. I was blessed to organize and lead the quorum with the help of the brilliant activist Carly Lettero.

AN ETHIC OF THE COSMOS

The story of the cold and frosty campsite is adapted from my essay "One Night, of Three Hundred Sixty-five," first published in *The Pine Island Paradox*.

The Buddhist ecophilosopher Joanna Macy is one of those who so beautifully call for three kinds of action. See her newly revised book with Chris Johnstone, *Active Hope: How to Face the Mess We're in without Going Crazy* (Gabriola Island, BC: New Society Publishers, 2014).

So many thanks to and so much admiration for Mary Evelyn Tucker. The book and film *Journey of the Universe* are sources for the ideas I share here, as is my aforementioned *Orion* interview "A Roaring Force from One Unknowable Moment." The interview is the source of the extensive quote on pages 123–124.

ETHICS AND EXTINCTION

The statistics about extinction come, again, from Carrington, "Earth Has Lost Half Its Wildlife in the Past 40 Years, Says WWF."

The paragraph about the language of extinction is adapted from my essay "The Parables of the Rats and Mice," in *The Pine Island Paradox*.

The argument against the language of the "Anthropocene" is adapted from my essay "Anthropocene Is the Wrong Word," *Earth Island Journal* (Spring 2013), 19–20.

Dave Foreman wrote his fierce answer in his essay "Wild Things for Their Own Sakes," in *Moral Ground*, 102.

"An Oath for the Wild Things" is adapted from my note "How can we preserve a future for grass sparrows and sea-grass, blue whales, krill, coral reef fish, lingonberries, grizzly bears—in fact, for all the extraordinary, beautiful, and tragically contingent variations of life on Earth?" in *Moral Ground*, 129.

The Portland writer David Oates sent me this wonderful riff on Noah's Ark by email on November 20, 2014. I reprint it with his permission. The full essay is David Oates, "Every Right Act is an Ark," *The Fourth River* (Spring 2016). *The Fourth River* is produced by the MFA program in Creative Writing at Chatham University in Pittsburgh, Pennsylvania.

My passage about environmental refugia was first published as "Refugia of the Toads," *Whole Terrain* 17 (2010), 46–47.

THE RIGHTS OF NATURE

This chapter is adapted from an online article I wrote for the Center for Humans and Nature, "Leaping Lizards! What Might It Mean to Recognize the Rights of Nature?" http://www.humansandnature.org/leaping-lizards—what-might-it-mean-to-recognize-the-rights-of-nature—article-175.php.

Christopher Stone's *Should Trees Have Standing?* was first published in 1972.

BREAKING THE SILENCE

The news of built-in global warming was announced by Coral Davenport in the *New York Times* article "Optimism Faces Grave Realities at Climate Talks," November 30, 2014, http://www.nytimes.com/2014/12/01/world/climate-talks.html?_r=0.

Lewis Mumford's words are from his paper "Authoritarian and Democratic Technics," *Technology and Culture* 5, no. 1 (1964), 1.

INVINCIBLE IGNORANCE

The Yale Project on Climate Change Communication is a valuable and reliable resource. The Web address is: http://environment.yale.edu/climate-communication/.

The explanation of the logic of climate denial is adapted from Michael P. Nelson and Kathleen Dean Moore, "Exposing the Logic of Climate Change Denial," *Oregonian*, December 1, 2012, http://www.oregonlive.com/opinion/index.ssf/2012/12/exposing_the_logic_of_climate.html.

Writer Mary DeMocker, who blogs as Climate Mom, published her argument about the New Deniers in the Portland *Oregonian*. "Climate-Conscious Oregonians Must Fight the New Deniers: Guest Opinion," *Oregonian*, March 15, 2014, http://www.oregonlive.com/opinion/index.ssf/2014/03/climate-conscious_oregonians_m.html.

FALSE PROMISES AND DEAD ENDS

Adaptation

My argument against an overemphasis on adaptation is adapted from my article "The Ethics of Adaptation to Global Warming," Center for Humans and Nature, http://www.humansandnature.org/earth-ethic-kathleen-dean-moore.

Scapegoating

Our guide in the H.J. Andrews Research Forest, where my students hooted for owls, was author and natural historian Tim Fox.

Warm thanks to Oregon State University philosopher and religion scholar Courtney S. Campbell, who helped develop and refine these arguments about Just War Theory. He is coauthor of "Owl v. Owl: Killing as a Conservation Strategy," the unpublished article from which this chapter is adapted. References to the authorities cited in this paper are:

Saint Augustine, "The Just War," in *The Political Writings of Saint Augustine*, ed. Henry Paolucci (Chicago: Regnery Gateway, 1962), 162–184.

Robin Brown et al., "Field Report: 2011 Pinniped Research and Management Activities at and below Bonneville Dam," October 2, 2011.

James F. Childress, *Moral Responsibility in Conflicts: Essays on Nonviolence, War and Conscience* (Baton Rouge: Louisiana State University Press, 1982).

Hannah Nevins and Michelle Hester, *Rakiura Tītī Restoration Project 2010 Annual Report*. Prepared for the Command Restoration Council, March 7, 2011.

U.S. Fish and Wildlife Service, *Revised Recovery Plan for the Northern Spotted Owl* (Strix occidentalis caurina), Portland, June 28, 2011.

Michael Walzer, *Just and Unjust Wars: A Moral Argument with Historical Illustrations*, 3rd ed. (New York: Basic Books, 2000).

Resilience

The argument about presilience draws on material in my essay "Presilience: Of Eggs and Baskets," *Connotations* (Fall 2012).

Despair

The incomparable Otis Redding wrote the lyrics to "(Sittin' on) The Dock of the Bay."

The beautiful charge by Clarissa Pinkola Estés is from a letter she called "Inspiration." I first found it on the Web at http://www.huna.org/html/cpestes.html.

Charles Goodrich, a wise friend and poet, sent me his wisdom by email, as did Hank Lentfer. Look for Charles's book of poems, *Insects of South Corvallis* (Corvallis, OR: Cloudbank Books, 2003), and Hank's wonderful memoir, *Faith of Cranes: Finding Hope and Family in Alaska* (Seattle: Mountaineers Books, 2011).

THE WORK OF SCIENCE

The *New York Times* article about James Hansen's retirement appeared on April 1, 2013.

Bob Dylan's song is "What Good Am I?" from the album *Oh Mercy*. You can see the full lyrics at his website: http://www.bobdylan.com/us/songs/what-good-am-i.

Be sure to read Kristin Shrader-Frechette, "An Apologia for Activism: Global Responsibility, Ethical Advocacy, and Environmental Problems," in *Ethics and Environmental Policy: Theory Meets Practice,* ed. Frederick Ferré and Peter Hartel (Athens: University of Georgia Press, 1994), from which these ideas are abstracted. In fact, you might want to read many more of her articles about the responsibilities of scientists and citizens. She's brilliant and bold.

Here's the reference to the letter to *Science*: Fakhri Bazzaz et al., "Ecological Science and the Human Predicament," Letters, *Science* 282 (1998).

THE WORK OF NATURE WRITERS

The conservationist Aldo Leopold wrote about the "wounded world" in *A Sand County Almanac* (New York: Ballantine Books, 1993), 165.

Robin Wall Kimmerer's latest book is *Braiding Sweetgrass: Indigenous Wisdom, Scientific Knowledge, and the Teachings of Plants* (Minneapolis: Milkweed Editions, 2013), with its beautiful intertwining of gratitude and responsibility.

I first issued the call to writers in the journal *ISLE: Interdisciplinary Studies in Literature and Environment* 21, no. 1 (2014), with journal editor Scott Slovic.

I wrote "It's a Bad Day for Rex Tillerson" for activist Mary DeMocker to read at the Eugene, Oregon, People's Climate March on September 21, 2014. Because of the miracles of instant communication, it was on that day shouted over the heads of marchers in a number of other cities, including New York.

I wrote the short piece "And Why You Must" for *Moral Ground,* where it appeared under the title "How Can We Express Our Love for the World?" I wanted to include it here, as it describes one of the few moments of true peace in an activist's life.

THE WORK OF WILDERNESS

This chapter is adapted from my article "A New Geography of Hope," *Earth Island Journal* 29, no. 3 (2014), 20.

WE HAVE MET THE ENEMY, AND IS HE US?

David Roberts makes his argument about our individual and collective responsibility in a *Grist* posting titled "How Climate Change Is Like Street Harassment," November 6, 2014, http://grist.org/climate-energy/how-climate-change-is-like-street-harassment/.

WHAT CAN ONE PERSON DO?

The answers to the quiz, which of course you all passed with flying colors: 1. (e), 2. (e), 3. (d).

A different, shorter form of "The Rules of Rivers" was printed in *Orion* 33, no. 4 (2014).

Carl Safina's quote in "Conscientious Objection" is from his essay "The Moral Climate," in *Moral Ground*, 325.

I first wrote about the student discussed in "Creative Disruption"—a student I still cannot name—for the feature article "National Defense," *Orion* 25, no. 2 (March/April 2006).

Michael P. Nelson and I worked together on the letter to grandparents, which appeared in *Moral Ground* (65–66)—but the hard-hitting postscript is all his.

Libby Roderick's song "Winter Wheat" is not yet officially recorded, but you can hear Libby singing it in a friend's living room: https://www.youtube.com/watch?v=8XiUe0NR2GU. Full lyrics are on her website. Lyrics reprinted with permission.

AFTER HOPE, THE ROAR OF THE LION, THE GREAT RISING WAVE

Gus Speth's book, from which his quotation comes, is *The Bridge at the Edge of the World* (New Haven, CT: Yale University Press, 2009).

More from the *New York Times* about the IPCC's newest alarm call is in Coral Davenport's aforementioned article, "Optimism Faces Grave Realities at Climate Talks."

See Macy and Johnstone, *Active Hope*. The material on framing is on page 167.

Derrick Jensen wrote against hope in "Beyond Hope," *Orion* 25, no. 3 (2006). The article begins, "The most common words I hear spoken by any environmentalists anywhere are, *We're f*cked*," and this was 2006. Michael P. Nelson made his argument against hope in his essay "To a Future without Hope," in *Moral Ground*, 458. I'm pleased to acknowledge their strong influence on my thinking.

AFTERWORD

My essay "Ring the Angelus," originally titled "The Call to Forgiveness at the End of the Day," is reprinted with permission from *Moral Ground*, 390–393.

ACKNOWLEDGMENTS

THE IDEAS IN this book have come from conversations and collaborations with many people. Foremost among these is philosopher Michael Nelson, my colleague at Oregon State University and collaborator on *Moral Ground*. I am pleased to thank him for the gifts that have come from our intellectual camaraderie over many years.

Long walks with scientists have been a source of pleasure and information. I'm especially grateful to hydrologist Gordon Grant, field biologist Molly Kemp, geologist Fred Swanson, ecologist Bruce Menge, botanist Robin Wall Kimmerer, climatologists of the Oregon Climate Change Research Institute, and the extraordinary Canadian aquatic ecologist Jonathan W. Moore, who is my son.

In the course of writing this book, I have asked for help from wonderful and generous writers. Kate Davis, Alison Hawthorne Deming, Charles Goodrich, Derrick Jensen, Carolyn Kremers, Hank Lentfer, Debbie Moderow, Gail Wells—thank you, dear friends. How wise you are, and how deeply you care about the world. Long conversations with other deep thinkers have shaped and informed my work: philosopher of religion Courtney Campbell, singer/songwriter Libby Roderick, concert pianist Rachelle McCabe, sociologist Mitchell Thomashow, historian of religion Mary Evelyn Tucker, and the extraordinary ecological architect Erin E. Moore, who is my daughter.

Special thanks go to the brilliant activists Mary DeMocker and Carly Lettero, who once were my students but are now my teachers. Thank you, Carol Mason, whose title is "writer's assistant" but who is this and so much more.

A number of nonprofit organizations have brought me into contact with brilliant people, new ideas, fertile and inspiring places—all of which fed this book. These include the National Park Service's Artist-in-Residence Program at Denali National Park, and especially Cass Ray; the Spring Creek Project for Ideas, Nature, and the Written Word, and especially director Charles Goodrich; the Island Institute in Sitka, and especially Carolyn Servid; the twelve women of the Mellon Foundation's Council on the Uncertain Human Future, especially Sarah Buie; the Center for Nature and Culture; *Orion* magazine; and the Whidbey Institute.

Most of all, I am deeply grateful to the man who has supported this work through the years with his insights, his time, his patience, his love, his computer savvy and sweep oars, his deep wells of information, and his faith in me and the choices I have made—the extraordinary biologist Frank Moore, who is my husband. Finally, love and gratitude to my grandchildren, Zoey Moore, Theo White-Moore, and Lem White-Moore, who mean the world to me.